Introduction

VISION IN ARITHMETIC AND ALGEBRA

THIS is a book about mathematics at a very low level. In recent years there seems to have been a sudden increase in the world's demand for mathematicians. A consequence of this has been that many people who do not feel themselves particularly qualified in mathematics find themselves called upon to teach it, either in an official or an informal capacity. A school's regular mathematics teacher leaves for a post in industry, and the French teacher has to take over his classes. A teacher of young children, who never liked arithmetic much and never did very well at it, finds that public attention is being focused on how she teaches it. Students at teacher training colleges have a similar experience. Parents find themselves unofficially involved in a similar dilemma. They want to help their children with mathematics but they fear that little good will come of the blind leading the blind.

LEARNING MATHEMATICS AND LEARNING CHINESE

The difficulty of learning a subject depends enormously on the way in which the subject is presented. A comparison may be helpful. Suppose that, instead of mathematics, we were setting out to teach Chinese. This seems a good comparison, because to many people mathematics seems much like Chinese. We will try to make the task sound as depressing as possible. To be a Chinese scholar you need to know 20,000 characters. In your first lesson you learn some of these. You go on until you know the lot. However depressing this may be for the pupil, the teacher's job seems straightforward enough. He must give the pupil a certain number of new characters each day to learn, and make sure the

1

Vision in Elementary Mathematics

pupil does not forget what he learnt on former days. The list for the first lesson might be something like Figure 1. The pupil begins to learn these 14 signs. On other days he adds others, and in due course reaches his 20,000.

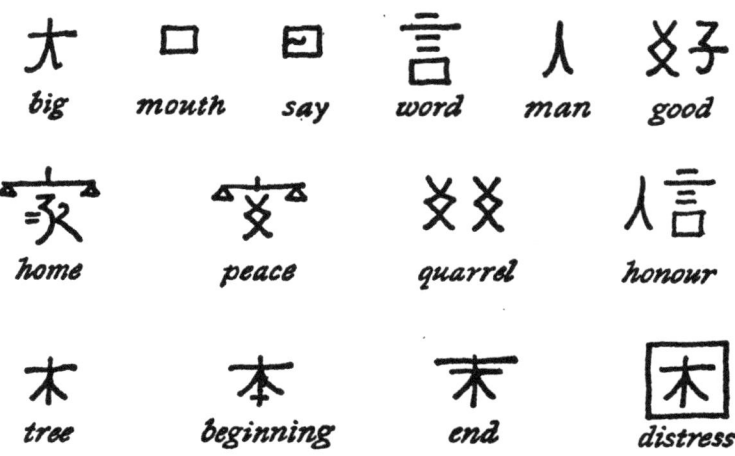

Figure 1

In this approach the pupil is, so to speak, outside the subject. There are these signs, something foreign to him, which he is required to memorize. But suppose the pupil is of an inquiring frame of mind. He begins to ask why these particular signs stand for these words. He puts himself in the position of the ancient Chinese scholars and wonders what he would do if he had to invent a written language with no precedents to help him. With concrete objects, of course, one could draw pictures. And indeed the character for 'man' does look rather like a man, at any rate it has two legs.* The character for 'mouth' looks very much like a mouth. For the verb 'say' a tongue seems to be wagging in the mouth, and 'word' shows hot air rising from the mouth. Great

---

* The explanations of the characters given here are true in their general spirit, but not necessarily in scholarly detail. It is, for instance, quite possible that the picture of the man arose from a side view, showing the body and an arm, rather than two legs.

Introduction

ingenuity has been displayed in the character for honour; it shows 'the man standing by his word'. For 'big', a fisherman holds out his arms to show the one that got away. The explanation of the character for 'home' is curious; it shows 'the pig under the roof'. In early agricultural society home was presumably the farm, the place where the animals were kept. There is also a roof in the character for 'peace'; peace is what you get when there is one woman under the roof. With two women, of course, a 'quarrel' is to be expected. The sign for a woman also appears in the character for 'good'. This character shows a woman and her baby – you can see the baby's head and its hands sticking out. The relation between a woman and her baby is one of 'love', hence 'lovely' or 'good'.

The last four characters are interesting. The character for 'tree' is a picture of a tree. A stroke drawn in the roots indicates 'beginning', very much as in the English sentence, 'The desire for money is the root of all evil'. (Exactly the same metaphor appears in the West African language *Twi*. One of the most remarkable signs of the kinship of all men is the way in which, in widely separated languages, the same concrete images are used to designate abstract ideas.) A stroke at the other end of the tree naturally represents the fruit, the final result, the 'end'. The character for 'distress' is perhaps the most remarkable of all. It shows a tree in a box, a living thing that wants to grow and develop, and is forcibly prevented. The men who, several thousand years ago, selected this out of all possible forms of distress, certainly had some insight into education. This character would make a good badge for a parent–teacher association.

It makes a considerable difference if a pupil is introduced to Chinese characters with some such explanation of how they arose. The pupil's feeling about the subject is improved by the knowledge that there is a reasonable explanation; he is not just learning a random collection of signs. His attitude becomes active instead of passive: he may begin to look through a Chinese dictionary and try to find other characters whose meaning he can explain. He may make guesses about the characters and try to test them. Needless to say, this approach does not remove all difficulties in learning to write Chinese. The simple ideas we have

3

Vision in Elementary Mathematics

seen are insufficient to explain 20,000 characters. But at least they give the pupil a good start.

One reason for the above discussion is to expose a widespread fallacy about teaching, the idea that remembering is easy and understanding difficult. John is a bright boy, we will teach him what the subject really means; Henry is dull, he will just have to learn things by heart. Now exactly the opposite is true; to remember things which you do not understand is extremely difficult. Ten years hence you may well remember that the Chinese character for 'quarrel' shows two women, because you understand the idea. You can see the reason for this choice of symbols. What you may very well forget is the exact shape of the character for 'woman'. For this is the part that depends purely on memory. No doubt this character originally was a picture of a woman, but in the course of the centuries it has become considerably simplified, and it does not look much like a woman now. So we cannot reason out how it ought to be; our memory has to work unaided. Of course, the character in Figure 2 is a very simple one, and

Figure 2

it is fairly easy to memorize. The difficulty perhaps appears more clearly if you try to draw the character for 'pig' which occurred in 'home', 'the pig under the roof'. This is a pure memory task, and, for most people, it is difficult. On the other hand, characters like 'big', 'mouth', 'say', and 'word' are extremely easy to remember. It is only necessary to run through them a few times, with the explanation of their origin in mind, and they become fixed in the memory. When you see them, the explanation comes readily to mind. Perhaps the reason why they are easier to learn is that we can concentrate on them for a longer time, because we are thinking about them. A meaningless sign we cannot think about. We can only look at it.

A very considerable part of the reputation mathematics has for being difficult is due to its being taught from the outside, as a memory task, rather than an exercise in understanding and in-

4

Introduction

sight. Of course, there are good teachers of mathematics, and it is unlikely that they will find anything very novel in this book. The aim of the book is simply to make methods of teaching mathematics, which are already used by many good teachers, known to a wider circle.

To mathematicians it should be made clear that this is not a book on mathematical philosophy or mathematical logic, neither of which will be discussed, but on mathematical vision.

CREATING CONFIDENCE

It is easiest to teach mathematics to very young children, for they have inquiring minds and they are self-reliant, and want to understand things for themselves.

It is much harder to teach adults, because so many adults have had their confidence shaken by bad teaching. They feel they are failures at mathematics, and we all shrink from attempting something at which we are likely to fail. This sense of embarrassment leads people to make remarks such as ,'Mathematics? Oh no, I could never do that at school. None of us could.' The object of this remark is to excuse oneself. I failed, but I was not to blame; you see, everyone fails; it is only human not to be able to do mathematics. Parents sometimes talk to their children like this, and they often fail to notice that, in trying to keep themselves on a pedestal, they are making it extremely likely that their children will fail in the same way. For a person who expects to fail does fail. A person who believes that only geniuses can learn mathematics will not learn mathematics. The proper thing for a parent to say is, 'I did badly at mathematics, but I had a very bad teacher. I wish I had had a good one.' This will have the advantage of being the true explanation. In mathematics, as in other subjects, with a good teacher the children enjoy the work.

The little digression on Chinese characters was meant to illustrate a technique for stimulating confidence. The general impression is that Chinese is very difficult, and so far as scholarly mastery is concerned this is probably true. The aim of the discussion earlier was to produce the reaction, 'Why, I understand this, I can see the meaning of these characters. Perhaps if I had

5

time I could learn Chinese after all.' It is that kind of reaction we want to produce in mathematics – 'Why, I understand this. I can see what it is about. Why didn't they tell me this at school?' Some of the material in Chapters 3 and 4 was designed with this aim. It has been tried on some parent–teacher associations and found to work fairly well. In fact, I have yet to meet anyone who could not understand the ideas explained at the beginning of these chapters.

The arrangement of the book presented some problems. A book that aimed at presenting mathematics in a very elementary way could conceivably begin with such problems as adding 3 and 2, and go on to more difficult questions such as 15+17. Now there certainly are great intellectual problems involved in the teaching of arithmetic to very young children. However, I could not myself get very excited about the question of 3+2. I know the answer is 5, and I expect many of my readers do too. I felt a book progressing steadily forward from the beginnings of arithmetic would be extremely dull. It seemed better to begin with questions a little further on and to roam around the subject in a series of narratives. The effect, it seems, is less dull, and yet in the end this book will be found to have dealt with many questions of elementary arithmetic, and I hope will be helpful to the teachers and parents of quite young children, as well as to those whose children are embarking on the beginnings of formal algebra.

Not knowing where this book may be used, I have felt free to be quite inconsistent. Sometimes discussions are presented as they might occur with a teacher and a class, sometimes as with a teacher or parent talking to a single child.

The whole book is based on the idea of the pupil discovering mathematics for himself.* As will be emphasized on more than one occasion, children often fail to solve problems because they cannot understand what the problems are. Naturally, if you do not know what a question means, you cannot be expected to give the correct answer. The main duty of a teacher is to show the pupil clearly what the problem is, and to encourage the pupil to attack

* 'Himself' of course is intended to cover 'herself'. There seems to be no satisfactory procedure in English for covering boys and girls by a single pronoun.

Introduction

it. Solving the problem is mainly the responsibility of the pupil. It is not the teacher's business to provide a cut-and-dried method of solution. This book is an attempt to show how such a philosophy works out in practice.

CHAPTER ONE

Even and Odd

SOME results in arithmetic can be grasped by a single act of mental vision. For example, suppose someone has played dice sufficiently to realize that the pattern ⚆ ⚆ ⚆ / ⚆ ⚆ ⚆ represents 6. It is immediately apparent that 6 consists of 2 rows of 3. In this sense, the result $6 = 2 \times 3$ is one that we can see directly.

Unfortunately, this kind of direct vision leaves off almost as soon as it has begun. In the lines above, the words *immediately, unfortunately, arithmetic* have occurred. How many letters are there in each of these words? Few people can answer without counting or breaking these words into smaller groups. One might perhaps see *arithmetic* as *arith metic* and realize that it contained 2 groups of 5 letters, making 10 in all. Even so, we have gone beyond the bounds of direct vision. We have a fairly clear picture of the smallest numbers. Beyond them, we see only a blur.

The average man is often too modest. He sees only a blur. He may blame himself for this; cleverer people, he may think, see clearly. But this is not so. The blur into which all numbers dissolve soon after 4 or 5 is the common experience of us all. If I am shown the answer to a multiplication question, such as 127×419, I do not know at a glance whether the answer is correct or not. I can sympathize with children who are helpless before the question 'What is 7×8?' I know, of course, that the answer is 56, because that was well drilled into me in my childhood, but it is not something that I can *see* directly. So it is understandable that children have no idea at all of how to cope with this question, and give all kinds of answers at random.

How to organize the chaos that lies beyond the smallest numbers is therefore a problem that confronts the entire human race.

While we cannot see the correct answer to 127×419, there are certain tests we can apply. We would, for example, reject immediately the answer 23 as being much too small, or 1,000,000,000 as being much too large. We would also reject the answer 53,312 although it is about the right size. For 127 and 419 are both odd numbers. It is impossible that their product should be an even number, 53,312.

The classification into even and odd brings us once more within the scope of vision. If we look at this number:

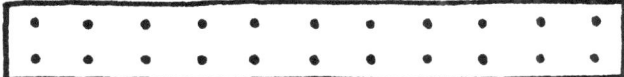

we can probably not tell without counting what number it is, but we can see at a glance that it is even. It has the characteristic shape of an even number. It can be divided into two equal parts, like this:

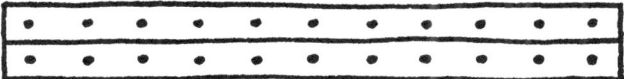

Also, it can be broken up into pairs, like this:

At a dance, if there are several couples on the floor, and no eccentric individuals dancing alone or in threes or other groupings, we can be sure that the number of people dancing is even. Evenness is thus a quality we can recognize in collections which we have not counted.

An odd number, on the other hand, has a shape such as:

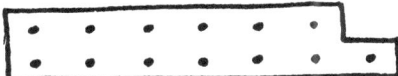

You cannot break it up into couples. One dot is left without a partner. Of course, we could place this shape the other way up:

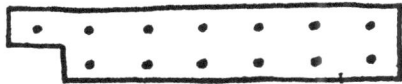

It would still represent an odd number.

9

Vision in Elementary Mathematics

The addition properties of even and odd are now apparent. Addition may be pictured as 'putting together'. If we put two even numbers together:

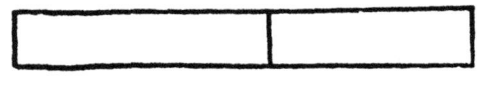

they form an even number, as we can see from the shape above. But if we put together an odd number and an even number, we obtain an odd number, as is shown by these shapes.

Finally, if we join together two odd numbers they dovetail to form the shape of an even number.

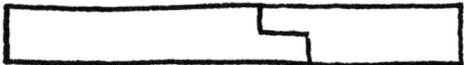

For multiplication, we might picture five sixes like this:

This has the shape of an even number. Clearly any number of sixes will give an even shape:

There is nothing special about six; the same illustration will serve for any even number. An even number multiplied any number of times gives an even number.

If we take an odd number and repeat it we obtain the following pictures.

An odd number, once ⬜⬛ **ODD**

An odd number, twice ⬜⬛⬜ **EVEN**

An odd number, three times ⬜⬛⬜⬛ **ODD**

An odd number, four times ⬜⬛⬜⬛⬜ **EVEN**

Even and Odd

Here it is fairly evident that odd and even shapes occur alternately. If we repeat an odd number an even number of times, the total is even. If we repeat an odd number an odd number of times, the total is odd. So 'even times odd is even, odd times odd is odd'.

PICTURES OF MULTIPLICATION

If you are seeking to understand, or to remember, or to teach mathematics through pictures, you should never allow yourself to be tied to one particular image of a mathematical idea. A picture that is suitable for one purpose may be most awkward for another.

In attempting to illustrate the result 'odd times odd is odd', I had to consider different ways of representing multiplication and see which was most suitable. I am by no means sure I chose the best.

Here are a few ways of picturing multiplication. We might, for example, picture 3 fours and 4 threes like this:

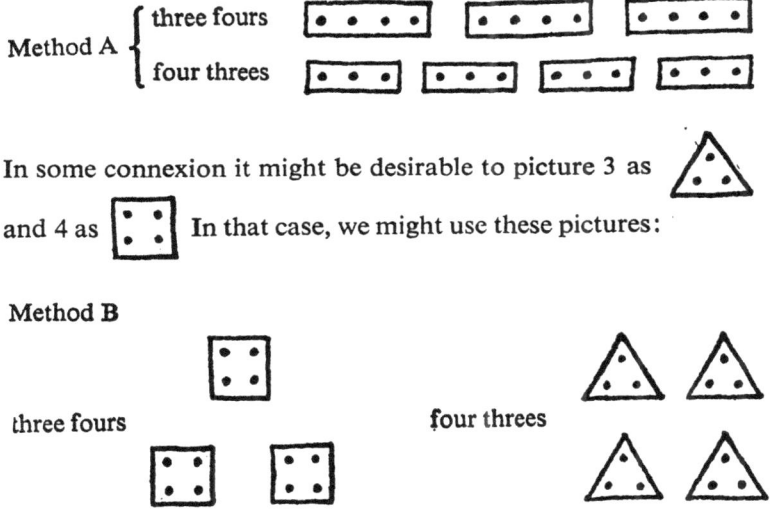

In both of these, the special patterns for 3 and 4 are apparent. None of the above brings out the fact that 3 fours and 4 threes

11

are the same number, 12. This can be shown by using a rectangle, as in Method C:

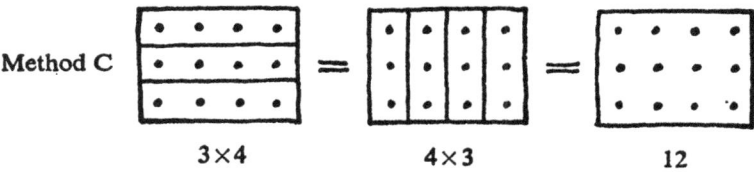

Method C

3×4 4×3 12

The rectangle is a very useful way of showing multiplication, and it will be used frequently in this book.

When we use the rectangle to illustrate multiplication, the numbers involved have to be represented by dots arranged in a straight line. For example, in the illustration of 4×3 above, the 4 is shown as 4 dots in a row, while the 3 is shown by 3 dots in a column. This arrangement does not emphasize the fact that 4 is even and 3 odd.

Method C is generally the most convenient way of picturing multiplication. However, it does not agree with the shapes we had earlier for even and odd. This means that there is a conflict when we want to portray simultaneously multiplication and the property of being even or odd. If we want to use Method C to indicate multiplication, we must find some new way of indicating even and odd. If we want to use our shapes for even and odd, we cannot use Method C to picture multiplication. Because of conflicts such as this, the pictorial representation of mathematical results calls for ingenuity and judgement. To devise such pictures, you need sustained meditation on the matter in hand, and this helps to fix the ideas in the memory. Learners should always be encouraged to make their own pictures. Even if the pictures are not good, the effort of making them will leave lasting traces in the mind, and cause the work to be remembered.

If we have 7 dots in line, and we want to emphasize that 7 is an odd number, we can do this by pairing the dots off, like this:

In the picture of 5×7 as a rectangle we may first stress the oddness

12

of 7, like this:

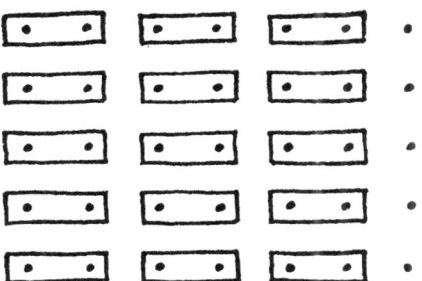

Unpaired dots now remain only in the last column, which contains, of course, 5 dots. Pairing these to emphasize the oddness of 5, we obtain the following picture of 5×7:

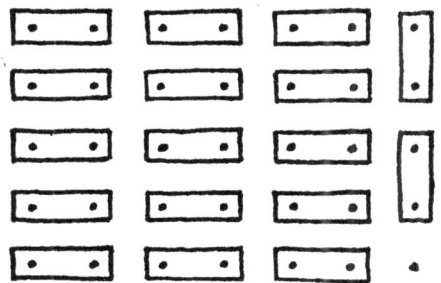

The solitary dot shows that 5×7 is in fact odd.

Here we have held to Method C for showing multiplication. If instead we decided it was more important to keep the shapes:

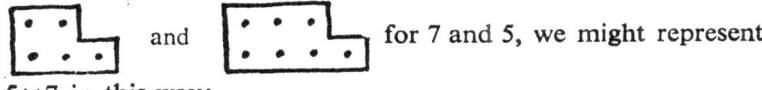 and for 7 and 5, we might represent 5×7 in this way:

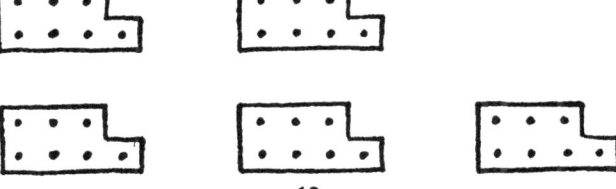

It might then occur to us that the oddness of this arrangement could be emphasized by turning the blocks like this:

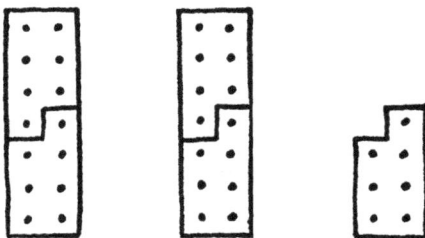

Again, a solitary unpaired dot stands out.

EVEN AND ODD IN ALGEBRA

If we place 2 fives side by side, we obtain the shape for an even number, 10.

There is nothing special about 5. If we took 2 sevens, or 2 fours, or 2 thirteens we should in each case obtain an even number. An even number results when we have 2 of any number you like to mention. We may picture this as here:

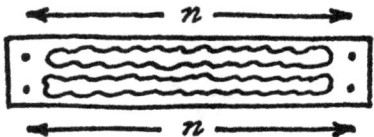

A cloud has come in front of the dots, so that I cannot see how many there are. The letter n is short for 'any number you like to mention'. As soon as you mention a number the clouds roll away and I see that number in the top row, and also in the bottom row. Whatever number you choose, the number of dots I see when the clouds lift will be even. I suppose this picture has a defect. If you were to choose 'zero' or 'one', my picture fails because I

14

Even and Odd

already have two dots in each row. I ask you to forgive this weakness. For any other number, the picture does what it is meant to. You may have doubts about what would happen if you chose 'two'. If you did that, the clouds would part and reveal a blank space.

Our picture thus helps us to imagine 'any even number'. In algebra the symbol corresponding to this picture is $2n$, which means 'twice any number you like to choose'.

The picture for an odd number will be:

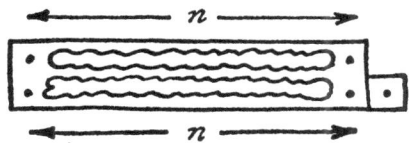

Here we see the number you chose twice, and in addition a single dot. Using n as short for 'the number you chose', this picture contains n twice and 1 extra. The symbol for it in algebra is $2n+1$.

We might make a little table to show the effect of choosing different values.

If n, the number you choose, is:	then $2n$ will be	and $2n+1$ will be
0	0	1
1	2	3
2	4	5
3	6	7
4	8	9

Every even number will appear, sooner or later, in the column under $2n$. Every odd number will appear, sooner or later, in the column under $2n+1$.

GENERALIZATION

An even number is a number that can be split into *two* equal parts. But *two* is no better than any other number. We might

15

equally well consider numbers that could be split into 3 equal parts, or for that matter into 5 or 11 or 17 equal parts. A similar theory could be developed for each.

If we are interested in splitting into 3 parts, we must recognize 3 types of numbers. There are numbers such as 15 which break exactly into 3 parts, and have the shape:

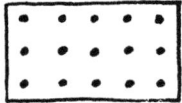

There are numbers such as 19 which leave a surplus of 1 when we try to split them into 3 equal pieces. These have the shape:

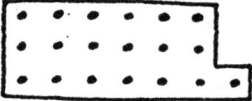

And finally there are numbers like 14 which leave a surplus of 2 and have the shape:

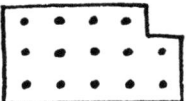

We have the words 'even' and 'odd' to describe how numbers stand in relation to 2. We have no words to describe how numbers stand in relation to 3. Rather than invent words, which would be a burden on the reader's memory, it seems wise to use the language of algebra, which gives us easily-remembered labels for these 3 types of number, as follows:

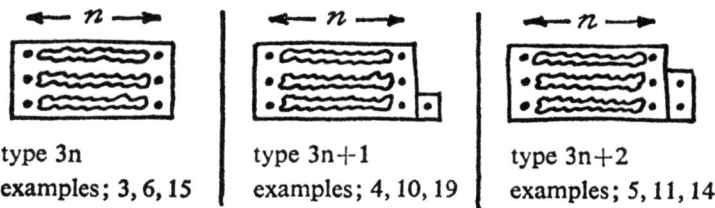

type 3n	type 3n+1	type 3n+2
examples; 3, 6, 15	examples; 4, 10, 19	examples; 5, 11, 14

16

Even and Odd

In the same way, a number would be placed in type $7n+5$ if, when you tried to split it into 7 equal parts, you found you had 5 dots left over. A number would be of type $10n+3$ if when you tried to split it into 10 equal parts you had 3 left over.

When we were dealing with even and odd numbers, we saw that we could predict the result of an addition or a multiplication if we knew the *types* to which the numbers belonged. For example, we could predict that an odd number multiplied by an even number would give an even number. We did not need to be told which odd number and which even number were involved in the multiplication.

This suggests a subject for investigation: what happens when we do not divide numbers into the types odd and even, based on 2, but into types based on other numbers. For example, if I ask you to write down any number that leaves remainder 3 on division by 5, and also any number that leaves remainder 4 on division by 5, and multiply together the numbers you have written, can I make any prediction about the answer you obtain? (I assume you carry out the multiplication correctly.)

In terms of our symbolism, I have asked you to multiply a number of type $5n+3$ by a number of type $5n+4$. Can we predict the type of your answer?

Someone with a working knowledge of algebra can deal with this question immediately. This does not mean that it is a trivial question. Many who have learnt algebra at school cannot do it. But for the moment we are not supposed to know anything about algebra. The question has been posed as one that might arise in the teaching of arithmetic. Two lines of attack are open. We might try to picture the multiplication. On page 13 we drew a picture to show that 5×7 must be odd. It is possible to adapt this idea and make it fit the present problem. Some readers may like this idea: others may not. It depends on their ingenuity and skill in devising diagrams. The other line of attack is to collect evidence by testing particular cases. This is straightforward, and could be done by children in an arithmetic class. They might choose 8 as a number of the type $5n+3$ and 9 as a number of the type $5n+4$. Then $8 \times 9 = 72$ and when 72 is divided by 5, the remainder is 2. So 72 is of type $5n+2$. Then they would select other numbers,

Vision in Elementary Mathematics

say 13 and 4. Then $13 \times 4 = 52$, which again is of type $5n+2$. The suspicion begins to grow that any number of type $5n+3$ multiplied by one of type $5n+4$ gives an answer of type $5n+2$.

Addition is much easier to handle than multiplication. From the pictures we had earlier, it is immediately evident that when you add together numbers of types $3n+1$ and $3n+2$ you obtain a number of type $3n$.

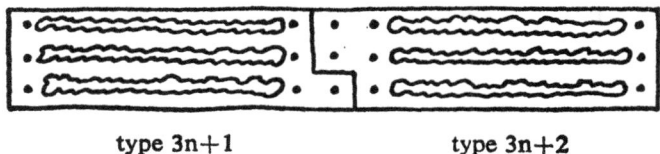

type 3n+1   type 3n+2

Any problem that involves only the *addition* of number types can be settled by drawing pictures. For example, it would not be hard to solve the question: if you add a number of type $7n+4$ to one of type $7n+6$, of what type will the answer be? As has been mentioned, it is possible to devise pictures to answer the question: 'What type of number do you get if you *multiply* a number of type $7n+4$ by one of type $7n+6$?' But it is simpler to deal with this by the methods of algebra, and we shall return to this question later on.

Exercises

1. Arrange 23 dots so as to show that 23 is of the type $3n+2$.
2. Draw a picture to illustrate a number of the type $4n+1$. Draw another picture to show a number of type $4n+2$. If the two pictures are joined together, what type of number do they represent?
3. We call a number of type $10n+3$ if on division by 10 it leaves remainder 3. Mention 5 different numbers of this type. If they are written, what do you notice about them?
4. Choose any number of the type $5n+2$ and any number of type $5n+3$. Multiply them together. What remainder is left when your answer is divided by 5? Repeat this work several times, with different numbers. Does the same remainder come each time?

18

Even and Odd

5. I multiply two numbers together. The first number is of the type $6n+2$. The second is of the type $6n+3$. What remainder does my answer leave when divided by 6?

6. If you multiply a number of type $3n+1$ by another number of the same type, what remainder do you get when you divide the answer by 3? To what type does the answer belong?

7. Choose two numbers, each of type $4n+1$. Multiply them together. Divide the result by 4. What remainder is there?

8. Choose two numbers, each of type $5n+1$. Multiply them together. Divide the result by 5. What remainder is there?

9. Questions (6), (7), (8) have something in common. Examine these questions and your answers to them. Do they suggest anything to you?

10. (An application of even and odd.) A football league contains 5 teams, whose names are given in the table below. The teams played only within the league. Do you find anything in the table below to suggest that an error has crept into the records?

	Won	Drew	Lost
Diehards	7	0	1
Erratics	6	1	1
Rovers	2	4	2
Wanderers	1	1	6
Invincibles	0	3	5

CHAPTER TWO

Divisibility

IN Chapter 1 we saw that 53,312 could not possibly equal 127×419, since 53,312 is an even number while 127×419 must be odd. This simple statement illustrates two ways in which the chaos of large numbers has become organized. The first way lies in the writing of the numbers themselves. If we were shown two heaps of stones, one containing 127, the other 128 stones, we could not tell which was which. But we distinguish with ease the written numbers 127 and 128. The second way in which organization has come is shown by the fact that we immediately recognize 53,312 as being an even number. This we certainly could not do if we had 53,312 actual objects before our eyes, lying in a disorderly heap. This chapter is concerned with these two themes – the writing of numbers and tests for divisibility. For when we recognize that a number is even, we are saying that it is divisible by 2. There are also ways of deciding quickly whether a number is divisible by 9, by 3, by 5, by 4, or by 8; if it is not exactly divisible, the remainder can be stated. The test for divisibility by 9 is often justified in algebra books. My impression is that many pupils learn to reproduce this proof without ever understanding it properly. This test, however, can be justified without any use of algebra by direct vision, and this we shall do. Further, the method used is applicable to the other numbers, 2, 4, 8, 5, 3, and even to 7. It is often said that there is no quick way of testing for divisibility by 7. In a way this is true. While our approach will give us a test for 7, it is certainly not a test of any practical value. However, it has some interest, not least the fact that the learner can develop it for himself.

THE WRITING OF NUMBERS

In Chapter 4 of Mark Twain's *Tom Sawyer* the system is explained by which the village Sunday school encouraged children to memorize passages from scripture. A child who could repeat two verses was given a blue ticket.

Divisibility

Ten blue tickets equalled a red one and could be exchanged for it; ten red tickets equalled a yellow one; for ten yellow tickets the superintendent gave a very plainly bound bible to the pupil.

The principle of this system is not hard to grasp. It is often not realized that an understanding of this principle, and of this principle alone, is sufficient for all computations, not merely with numbers, but with pounds, shillings, and pence or with weights and measures of any kind. It is truly remarkable that so many schools feel it necessary to drill their students in money sums and in weights and measures for months and years, when the intellectual content of these topics is so slight. The custom of doing this is no doubt a hangover from the nineteenth century, when arithmetic was still done by men rather than machines. Even then, much of what was taught could have been done more easily with the help of a ready reckoner – compound interest, for example. In the age of adding machines and electronic computers the drilling of children in routine calculations with money and measures is a fantastic waste of time. The practical needs of life, no doubt, require everyone to be familiar with money, weights, and measures. Beyond that an approximation is usually sufficient. I know that a mile is rather more than 5,000 feet. If in an aeroplane I hear an announcement that we are 10,000 feet up, I may want to know that our height is roughly 2 miles. I do not want to know that it is 1 mile 7 furlongs 1 chain 11 yards 1 foot. It is good for children to acquire a general feeling for how large things are. But, above all, it is necessary for children to understand what is involved in a calculation. If they understand what is happening they can devise methods for themselves and can test by their own thinking the correctness of their work.

Imagine then a pupil preparing to attack a problem in weights and measures, or a money sum. We suppose this pupil to understand the basic ideas – that addition means putting together, subtraction taking away from, and so on for multiplication and division. In addition, he understands the idea on which Mark Twain's Sunday school operated – that several objects of low value may be exchanged for one of higher value. Now suppose him confronted with a currency system and a way of writing money that is as novel to him as it is to us. The only coins and notes that

Vision in Elementary Mathematics

this currency employs are the penny, sixpence, half-crown, ten-shilling note, and pound note. All of these are used in describing sums of money. Thus £5/1/3/4/2 would mean 5 pounds, 1 ten-shilling note, 3 half-crowns, 4 sixpences, and 2 pennies. In order to work with this system all that we, or the pupil, would need to understand would be the values of the units; to paraphrase Mark Twain, 6 pennies equal a sixpence and could be exchanged for it; 5 sixpences equal a half-crown and could be exchanged for it; 4 half-crowns equal a ten-shilling note; 2 ten-shilling notes equal a pound. To avoid any effort of memory, suppose these facts were written down, or displayed by a poster on the wall. Suppose now we wish to find the sum of £5/1/3/4/2 and £2/1/2/4/5. The system is unfamiliar to us. We are starting almost like children, so let us do it as it might be done in primary school. We set out the actual objects involved – 5 pound notes, 1 ten-shilling note, 3 half-crowns, and so on, representing the first amount, and another collection representing the second amount. We have to add these, that is, we have to put them together. We merge the two collections. It is now apparent that some exchanging is in order. We have 7 pennies. It will be good to exchange these for 1 sixpence and 1 penny. There are now 9 sixpences on the table. It is in order to exchange 5 of these for a half-crown. After doing this, there will be 6 half-crowns on the table, 4 of which can be exchanged for a ten-shilling note. And so we continue, and in due course have on the table 8 pounds, 1 ten-shilling note, 2 half-crowns, 4 sixpences, and 1 penny. This we record as £8/1/2/4/1, the answer to the question.

In view of the difficulty so many people find with arithmetic, it seems worth while to examine the procedure outlined above, and see what it involves. It seems reasonable to believe that this procedure could be taught even to persons of very limited intelligence. Bear in mind that the objective is, at present, a limited one. We are not concerned with paper calculations. The learner is supposed to have the actual money in front of him, and to use the coins and notes as apparatus for arriving at the answer. The question is given in written form, and uses an unfamiliar system. The answer has to be written in the same system. What must a person know in order to carry out the task correctly? He must, of

22

course, be able to count and understand how to write small numbers. (The largest number written in the work above is 8.) He must associate the word 'add' and the sign ' + ' with the action of putting two collections of objects together. He must understand the idea of exchanging, and be able to use correctly the table provided, which tells him how many pennies may be exchanged for sixpence, how many sixpences for a half-crown, and so forth.

We suppose a learner carries out tasks of this kind – always with the actual material – until he is thoroughly familiar with the procedure. With a learner of very high intelligence, two or three repetitions might be sufficient; with a slow learner, hundreds or even thousands might be required. But quickly or slowly, the stage is reached at which the procedure has been thoroughly imprinted on the learner's mind. He can now imagine it without actually doing it. Of course, imaginations vary immensely; some are vivid and detailed; some are obscure and dim. A good teacher will try to discover by conversation what each pupil can imagine, and will be careful not to discourage the pupil by demanding more of his imagination than it is able to provide.

Quickly or slowly, then, the learners come to the stage where they understand the procedure. They are then ready for the second stage, in which the actual physical operations are replaced by mental images or memories of those operations. The learner describes the process to the teacher. 'First, I put the 2 pennies and the 5 pennies together. That gave me 7 pennies. Then I changed 6 of the pennies for a sixpence. That left me with 1 penny and an extra sixpence, which I put with the 8 sixpences there already.' And so it continues. The words mean something to the student, because he has many times performed the actions. The teacher may indicate a convenient way of recording the process. For example, the beginning of the operation, as described by the pupil earlier in this paragraph, might be recorded as in Figure 3.

The work is now beginning to look very much like conventional arithmetic, and someone might ask, 'Wouldn't you get just the same result if you told the child to put down 1 and carry 1?' Certainly – if you were lucky – you might, by giving such instructions, get the child to write down the correct figures. But the

23

Vision in Elementary Mathematics

child would not have acquired understanding: he would merely have learnt a rule. And by the time he had learnt several such rules, he would begin to forget when each rule should be used, and he would start to write nonsense. By the approach we have outlined, there is no strain on the memory at all. If the pupil spent several years without doing any work on weights and measures, he could rapidly think his way through any problem that arose.

pounds	ten shilling notes	half crowns	sixpences	pennies
5	1	3	4	2
2	1	2	4	5
			1	

1

Figure 3

Anyone who is likely to be consulted by a child on questions of arithmetic will find it worth while to think through the physical procedure by which problems in arithmetic can be solved. A child may be puzzled by a standard process such as simple division. Many adults cannot give at all a convincing explanation of it. By stating a question in terms of real objects, the problem becomes dramatized and easier to grasp. Suppose, for example, in English currency (as usually handled, not the fiction I used earlier) it is necessary to divide £1 equally among 6 persons. You can explain the steps to a child: first exchange the pound note for 20 shillings; then give 3 shillings to each of the 6 persons; exchange the remaining 2 shillings for 24 pennies, and then give 4 pennies to each person. None of these steps is mysterious. The child should be able to follow the process, and the written work follows and records the physical procedure. Of course, in doing written work the child may stumble on some detail; he may forget, for example, what 24 divided by 6 is. But that is a separate question; it calls for a separate discussion, and for its own physical demonstration.

Exactly the same ideas are involved when we are dealing with pure numbers instead of money, weights, and measures. The work is indeed simpler, since we can forget the arbitrary numbers – 12 inches = 1 foot, 3 feet = 1 yard – and remember the single number '10'. Mark Twain's Sunday school was already a dramatization of our usual system – 10 blues for 1 red, 10 reds for 1 yellow, 10 yellows for a bible. The number we write as 2,345, is the number of blue tickets equivalent to 2 bibles, 3 yellow, 4 red, and 5 blue tickets. Just as we dramatized the sharing of £1 among 6 people, we could dramatize the sharing of 1 yellow ticket among 4 scholars who had earned it by a cooperative effort. The yellow is converted to reds; the reds are shared out as far as possible; any surplus reds are exchanged for blues, and so on. The procedure follows exactly that of the conventional division process for $4\overline{)100}$.

The role that 10 plays in our arithmetic is no doubt due to the fact that we have 10 fingers (thumbs counting as fingers for this purpose). The role of 10 is a biological accident: 10 has no mathematical properties to make it superior to the other numbers. If mankind had only 8 fingers, it is almost certain that the village Sunday school would have given 8 blues for each red, 8 reds for each yellow, and so on. On this basis, a system of writing numbers could be built, 234 standing for the number of blue tickets that were equivalent to 2 yellow, 3 red, and 4 blue tickets. This system of writing numbers would be different from ours. Their sign 234 would stand for quite a different number from our 234. But they would stay inside their own system, and it would work just as well as ours.

The results, of course, would look strange to us. In their arithmetic, $6+6 = 14$ would be perfectly correct, for their 14 signifies 1 red and 4 blue. With the value of a red ticket fixed at 8 blue tickets, 1 red and 4 blue equals 12 blue. In order to avoid confusion between their system and ours, it is usual to write a small 8 after their signs. Thus 14_8 means 'the number that would be written 14 by a race with 8 fingers'.

The idea that the writing of numbers need not be based on 10 is not a new one. How old it is I do not know. It was certainly current in the 1880s; Hall and Knight's *School Algebra* and *Higher Algebra* both contain chapters on this subject. In the 1940s, Leicester Teacher Training College required their students to

Vision in Elementary Mathematics

work out how the multiplication tables and other parts of arithmetic would have looked if we had possessed 8 fingers instead of 10. Recently the study of 'bases other than 10' became very popular in the United States. The use of the binary scale in electronic computers helped to bring it back into fashion.

The great value of this topic for teachers is that it makes us children again. When we work with our usual arithmetic, we can remember the answers. If a child forgets 7×6 we think, '42 of course. What else could it be?' But in the scale of 8, it is not true that $7 \times 6 = 42_8$. For 42_8 means 4 eights and 2 – what we call 34. We have to think out for ourselves what 7×6 is in this unfamiliar system, and it teaches us both arithmetic and patience with children.

Also, these other systems give us a wealth of material to explore. For example, if we had 7 fingers, 10 would be written 13_7 and 16 would be 22_7. Both of these are even numbers. In our system we recognize an even number by looking at the final digit. This test clearly fails for the 7-finger arithmetic. Ten is written 13_7; the final digit is 3, which is odd; yet 10 is certainly an even number.

There are many other questions that arise when we compare our system with others, and it can be quite interesting to experiment and explore.

The simplest of all is the binary system in which grouping is always by pairs – 2 blues for a red, 2 reds for a yellow, and so on. We might indicate the basic tickets like this:

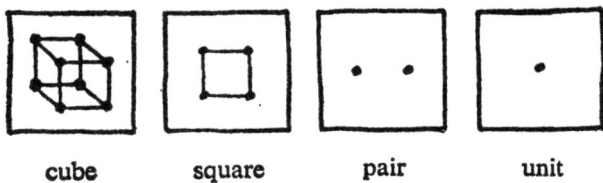

| cube | square | pair | unit |

Of course, there are more tickets for higher numbers. (I do not have any good names by which to call the tickets with 16 and 32 dots.) By selecting some or all of the tickets shown above, we can obtain any number of dots up to 15. 11, for example, would be shown as:

26

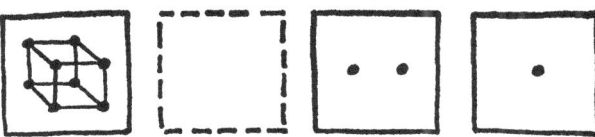

It contains 1 cube, 0 square, 1 pair, and 1 unit. Accordingly it would be written 1011_2.

Children enjoy deciding which cards have to be chosen to give any number, and recording the result in binary notation.

Addition in binary (or in any other scale) does not involve any new principle. It can be thought out just as was done with the fictitious currency.

Equally, of course, the operations in our usual system, based on 10, can be thought out. Instead of pair, square, cube we now have 10, 100, 1,000. The Montessori system uses a single bead to represent a unit. To represent 10, 10 beads are threaded on to a straight wire. For 100, 10 of these wires are placed side by side to form a square. For 1,000, 10 squares are piled on top of each other to form a cube. So we reach the representation in Figure 4.

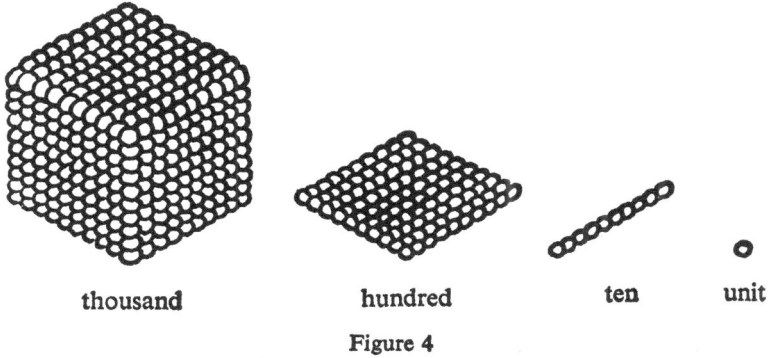

thousand   hundred   ten   unit

Figure 4

It is necessary to emphasize that the cube is full of beads. Children are liable to think of the beads on the surface – the beads they can see – as the ones we are speaking about. This is particularly liable to occur if a picture is drawn instead of the actual material being handled.

27

Vision in Elementary Mathematics

This apparatus, or some similar device, should be available when children are learning arithmetic. And this does not apply only to young children. Quite old boys and girls when they write 243 often are not aware of the meaning of what they write. They do not have before them the vision of the arrangement in Figure 5

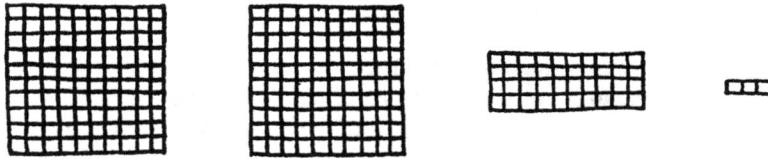

Figure 5

It would, of course, be ridiculous to make older boys and girls do all their work with sets of beads. With older children, the apparatus should be used mainly when some difficulty has arisen. If the children have been well taught in their early years, the apparatus reminds them of the meaning of symbols and assists the imagination. If the children have been rote-taught in their early years, the apparatus shows them for the first time what arithmetic is all about. The apparatus will then have to be used quite frequently. We are making up for what they should have had in the kindergarten, but missed. The relevant question is not how many years the pupils have been alive; but to what experiences they have been exposed in those years. In a course for adults it would still be in order to use apparatus. Things convey a message that words fail to do.

TESTS FOR DIVISIBILITY

The use of apparatus can be well illustrated by explaining the test for divisibility by 9.

This test is well known. Suppose, for example, we want to find what remainder, if any, 2,365 leaves on division by 9. We add together the digits that occur in 2,365. This gives $2+3+6+5=16$. We divide 16 by 9. The remainder is 7. This means that 7 will be the remainder when the original number 2,365 is divided by 9. You can check this by carrying out the actual division.

Divisibility

Why does this method work? We try to dramatize the ideas involved. First, we have the number 2,365, which corresponds to the collection of Montessori beads shown in Figure 6.

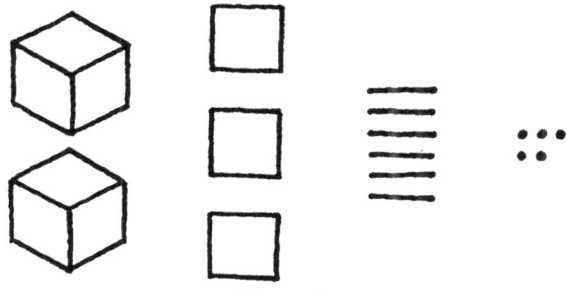

Figure 6

We have to divide this number by 9; that is, we have to share out this collection of beads among 9 people, who insist on being treated alike.

Since the number is built of units, 10s, 100s, and 1,000s, we make a preliminary study of these.

First, suppose we had a single unit. We cannot give it to any of the claimants. We make no distribution of beads at all. We have this single bead left on our hands.

Next, suppose we have a 10:

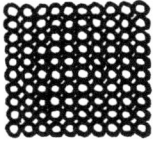

We can give one bead to each of the claimants, and again we have one bead left on our hands.

We now consider what we should do with 100 beads:

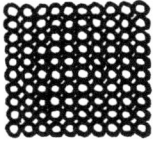

We can break it up into 10 sticks of 10. We hand out a stick to each person, and that leaves us with a stick of 10 to be shared. But we have already thought what happens when 10 beads are shared out. Each person gets 1, and we are left with 1.

29

When we come to 1,000:

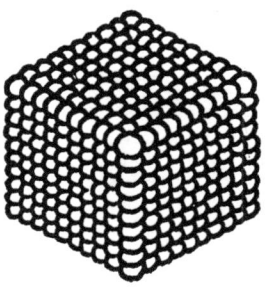

we naturally break this cube into 10 hundreds, give each person 100 and then think how to share out the remaining 100. But we know from the earlier work what happens when we share out 100. We get left with 1 bead on our hands. (We are only interested in the remainder. We are not at present concerned with how many each person gets.)

So, if you have to share out a single unit, or 10, or 100, or 1,000, you are in each case left with a remainder of 1. It should be clear that this argument goes on for ever. If you had to share out 10,000, or 100,000, or a million, you would in each case be left with one bead on your hands.

Now we come back to our picture of 2,365 beads. We share this amount out among 9 people, but not by the usual method of arithmetic.

We begin with the thousands (the cubes). We know that 1,000 can be shared out equally among 9 people, and leaves 1 bead over. We see 2,000, so we share each 1,000 out fairly, and have 2 beads left over at the end – 1 from each 1,000.

In the same way we deal with the hundreds. Each 100 can be shared out, with 1 bead left over. There are 300. So sharing out the hundreds leaves us with three beads, 1 from each 100.

In the same way, we share out the tens, and have 6 beads left over, 1 from each 10.

The 5 individual beads remain on our hands.

At this stage, each of the 9 people has received the same number of beads. Each 1,000 has been shared out fairly, each 100, each 10.

A certain number of beads remain (see Figure 7).

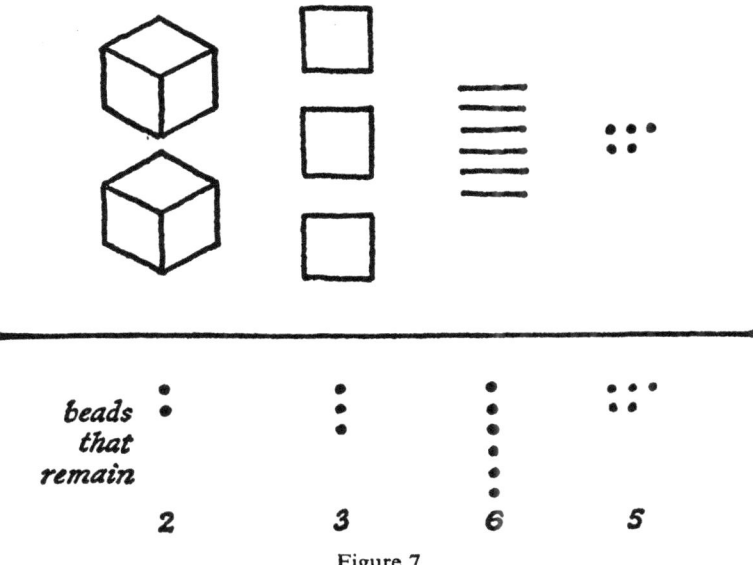

beads
that
remain

2   3   6   5

Figure 7

The number of beads left on our hands, at this stage of the process, is $2+3+6+5$. Since $2+3+6+5=16$, which is more than 9, we can give a further bead to each person, and we are left with 7 beads. We can do no more; 7 is the remainder.

You will see that the number $2+3+6+5$ arises quite naturally. At a certain stage of the sharing out, the number on our hands has been reduced from the 2 thousands, 3 hundreds, 6 tens, and 5 to the much smaller number $2+3+6+5$. What remains when we share out this smaller number tells us the remainder for the original larger number.

We have chosen particular numbers, 2, 3, 6, and 5. But the argument does not depend on this particular choice. There is nothing special about these. Anyone who understands this has in fact passed from arithmetic to algebra. Arithmetic studies what particular numbers do, algebra what any numbers do.

If in our argument we replace 2 by any number a, 3 by any number b, 6 by any number c, and 5 by any number d, we arrive at the conclusion, 'When a thousands, b hundreds, c tens, and d units is divided by 9, the remainder is the same as when $a+b+c$ $+d$ is divided by 9.' One might explain the use of the letters a, b, c, d to children as follows. Alf is going to be allowed to say which number a stands for. Bert may choose what number b is to represent. Charlie may choose a number for c and Dave for d. Whatever numbers the boys choose, the statement will be found to be true.

NUMBERS OTHER THAN NINE

The test for dividing by 9 is well known, and it is useful in arithmetic lessons. Many children, for example, know that 54 and 56 have something to do with 9×6 and 7×8, but they do not know which is which. Our test allows us to spot a number in the 9 times table at once. We see that 56 cannot be in the 9 times table, for $5+6=11$ which does not divide exactly by 9. (In fact, our test shows that it is 2 more than a multiple of 9.) On the other hand, 54 gives us $5+4=9$, so 54 is a multiple of 9.

It seems to have been overlooked that the same type of test can be devised for numbers other than 9. Suppose, for example, we continue our discussion of 54 and 56. We want to test whether 54 is in the 8 times table; 54 means 5 tens and 4. What happens if we try to share 10 beads among 8 people? We have 2 beads left over. If we follow the same procedure with 8 that we did earlier with 9 – that is, if we share each 10 out separately – each 10 will leave us with a pair of beads. We shall reach the picture in Figure 8.

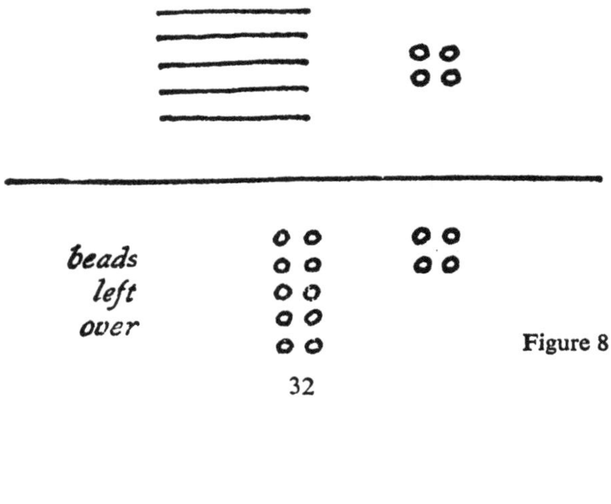

Figure 8

Divisibility

By distributing 8 beads from each stick of 10, we reduce the 5 tens to 5 pairs. Thus 5 tens and 4 units are reduced to 5 twos and 4 units. Since 5 twos and 4 make 14, which is not in the 8 times table, the original number 54 is not in the 8 times table. When the same test is applied to 56, we find 5 twos and 6 makes 16, which is in the 8 times table, so the original number 56 is also in the 8 times table – as it should be.

Almost everyone knows the pattern of the 9 times table.

$$18 \text{ with } 1+8=9$$
$$27 \text{ with } 2+7=9$$
$$36 \text{ with } 3+6=9, \text{ etc.}$$

The corresponding pattern for the 8 times table is this:

$$16 \text{ with } 2\times1+6=8$$
$$24 \text{ with } 2\times2+4=8$$
$$32 \text{ with } 2\times3+2=8$$
$$40 \text{ with } 2\times4+0=8$$
$$48 \text{ with } 2\times4+8=16$$
$$56 \text{ with } 2\times5+6=16$$
$$64 \text{ with } 2\times6+4=16$$
$$72 \text{ with } 2\times7+2=16$$
$$80 \text{ with } 2\times8+0=16$$

In the language of algebra, if the number consisting of a tens and b units is in the 8 times table, then so is the number $2a+b$.

Question for investigation. What is the corresponding property of the 7 times table?

The discussion of divisibility by 8 has so far been confined to numbers less than 100, which is quite sufficient if we simply want some way of bringing variety into the teaching of the multiplication table. But, just as with 9, a test for divisibility by 8 can be found that applies to numbers of any size. The test for 9 was discovered by a study of the basic numbers 1, 10, 100, 1,000 and so on. Exactly the same procedure may be used for 8 (or for any other number).

It would be possible simply to divide 1, 10, 100 etc. in turn by 8, and to record the remainders. This would give us the following information.

Number	10,000	1,000	100	10	1
Remainder when divided by 8	0	0	4	2	1

The noughts are a new feature. The 0 under 1,000 means that 1,000 beads can be shared out exactly among 8 people. Once the noughts have begun to appear they are bound to continue. We can be sure that 0 will appear as the remainder for 100,000, for a million, for ten million and so on. For if a number can be shared out exactly among 8 people, so can 10 times that number.

Accordingly, to find the remainder, we do not need to bother about the larger end of our number. The millions, the hundred thousands, the tens of thousands, the thousands – all of these can be shared out exactly. They go and leave no trace. But each time we share out a hundred, 4 beads are left over. Each 10 that is shared out leaves us with 2; and, of course, each unit leaves us with 1.

For example, if we are engaged in sharing out 729,356 beads by this method, we need not bother about the 729,000. This consists of complete thousands and can be shared out without remainder. The 3 hundreds will leave 3×4 beads over; the 5 tens will leave 5×2 over; the 6 units will leave 6. As $3 \times 4 + 5 \times 2 + 6 = 28$, we can distribute 3 more beads to each person, and the final remainder will be 4.

Thus, to find the remainder on division by 8, we add 4 times the number of hundreds and twice the number of tens to the number of units. We see what remainder this total leaves on division by 8. The original number leaves the same remainder.

In algebraic notation, if the number contains h hundreds, t tens, and u units, we consider $4h + 2t + u$. This number and the original number have the same remainder on division by 8.

The test for divisibility by 9 uses all the digits: the test for 8 uses merely the last 3 digits.

Only the last 2 digits occur in the test for divisibility by 4. For 100 is exactly divisible by 4. So the hundreds, thousands, tens of thousands, and so on leave no trace when we are dividing by 4. But 10 leaves a remainder of 2. Thus, however large the number,

Divisibility

we can test divisibility by 4 by adding twice the tens digit to the units digit. If we want to know what remainder 1,965 leaves on division by 4, we ignore the 1 and the 9. Twice 6 added to 5 gives 17, which is 1 more than a multiple of 4. The remainder is 1.

The extreme case is the test for divisibility by 2. 10 is an even number; 10 objects can be shared fairly between 2 people. 100 can be broken into tens, so it also must be even. The same applies to 1,000. So, in sharing out among 2 people, the thousands, hundreds, and tens go and leave no trace. By looking at the units figure alone, we can see what remainder the number leaves when we divide by 2.

The same argument shows why we need only look at the last digit to see whether a number is divisible by 5.

The test for divisibility by 3 resembles the test for 9, in that we have to take account of all the digits. It does not raise any new questions of principle. The rule can be thought out by the methods already discussed – the study first of 1, 10, 100, 1,000, and so on; then the 'dramatized' sharing out of beads among 3 people.

Whether a number is divisible by 6 is usually settled by checking first whether it is even, and then whether it is divisible by 3. If it passes both tests, it must be divisible by 6.

A direct test for divisibility by 6 can be obtained, quite straightforwardly, by following the procedure used throughout this section. It may amuse you to find this test, which however is not of much practical value.

The next section applies the same procedure to test divisibility by 7. Before reading on, you may wish to try your hand at doing this for yourself.

DIVISIBILITY BY SEVEN

We begin as usual by considering 1, 10, 100, 1,000, and so on. The remainders when these numbers are divided by 7 are as follows:

Number	1,000,000	100,000	10,000	1,000	100	10	1
Remainder	1	5	4	6	2	3	1

The most obvious, though not the most economical, way of obtaining these remainders is to work each out separately. Ways of reducing the labour will be considered later.

Vision in Elementary Mathematics

Suppose we wish to know the remainder when 4,325,643 is divided by 7. We follow our usual procedure. We divide out each million, then each 100,000, and so on. Each million leaves a remainder of 1, so 4 millions leave 4×1. Each 100,000 leaves 5, so the 3 hundred thousands leave 3×5. In the same way, the 2 ten thousands leave 2×4; the 5 thousands leave 5×6, the 6 hundreds leave 6×2, the 4 tens leave 4×3 and the 3 units leave 3×1. So there are altogether $4+15+8+30+12+12+3=84$ beads left over. As 84 divides exactly by 7, the original number divides exactly by 7. The test, as was admitted earlier, is far from being of practical value!

If you investigate what happens when still larger numbers – 10,000,000; 100,000,000; etc. – are divided by 7, you will find that the sequence 1, 3, 2, 6, 4, 5 repeats itself indefinitely.

PROPERTIES OF THE REMAINDERS

Earlier, when we were considering division by 8, we had these results:

Number	1,000	100	10	1
Remainder	0	4	2	1

Here it is very noticeable, as we read the numbers 1, 2, 4, that each is twice the previous number. If we continued doubling, the next number would be 8. But we cannot have remainder 8 when we are dividing by 8, so this 8 is replaced by 0.

Why does this doubling effect occur? When 10 is divided by 8, there is a remainder of 2. When we come to share 100 beads among 8 people, we might do it as follows. 100 is 10 tens. We give out 8 tens, and are left with 2 tens. These tens when shared out each leave a remainder of 2. So altogether we have 2×2 beads left. When we come to share out 1,000 beads, since 1,000 is 10 hundreds, we hand out 8 hundreds and are left with 2 hundreds. So we might expect the remainder from 1,000 to be twice the remainder from 100. However, as twice 4 is 8, we are able to hand out these 8 beads, and be left without any remainder.

Similar considerations apply whatever number we are dividing by. When we are dividing by 7, instead of doubling we multiply

36

Divisibility

by 3, since 3 is the remainder when 10 is divided by 7. For example, when sharing out 100 beads, we break 100 into 10 tens; we hand out 7 tens, and are left with 3 tens. Our first impression might therefore be that the remainder from 100 would be 3 times the remainder from 10. But that would give 9; we have to give out 7 beads, and the remainder is reduced to 2. In going from 100, with remainder 2, to 1,000, with remainder 6, we see that the remainder has, in fact, been multiplied by 3. When we go on to the next stage, we multiply 6 by 3. This gives 18; 2 sevens can be taken out of this, so that the number is reduced to 4, the correct remainder for 10,000. Continuing in the same way, $3 \times 4 = 12$; remove 7, and we have 5, the remainder for 100,000. 3 times this remainder is 15; 2 sevens have to be removed, and we obtain 1 as the remainder for a million.

The same numbers 1, 3, 2, 6, 4, 5, occur as remainders when we work out the decimal for 1/7.

Since a million leaves remainder 1 when divided by 7, the number 999,999 must be exactly divisible by 7. In fact, $999,999 \div 7 = 142,857$. This number has a certain 'trick' property. If you multiply it by 2, 3, 4, 5, or 6, you get a striking result. For example, $142,857 \times 3 = 428,571$. The 1 has been shifted from the beginning to the end.

Plenty of investigations can be based on the ideas discussed in this chapter. If we all had 7 fingers, how would we count? How would we write 100? What would the addition and multiplication tables look like? Would there be a table that behaved as the 9 times table does in our usual arithmetic? How would we test for divisibility by 4? A learner who became very enthusiastic might work systematically through the various systems – the 2-finger system (binary), the 3-finger system, the 4-finger system, and so on, observe any interesting features of these, and devise tests for

Figure 9

37

divisibility in each system. The questions below are only samples, to show the possibilities.

Reminder: 123_5 indicates that a number has been recorded in the 5-finger system; it can be pictured as the collection of dots shown in Figure 9.

Exercises

1. How would our number 'ten' be written in the 5-finger system? How do we normally write the number 14_5?
2. How is 'ten' written in the 7-finger system? How do we normally write 42_7?
3. Write the numbers 1 to 10 in the binary system (the 2-finger system).
4. In the binary system, how would you write the number that is 1 more than $1,111_2$?
5. In the 3-finger system, what is the largest number that can be written with 3 digits? (In answering, show both how the number is written in the 3-finger system, and how it is written in our usual system.)
6. How do we usually write 141_5? Is this number even or odd?
7. Which of the following numbers are even, and which are odd?
$$1_3; \quad 11_3; \quad 111_3; \quad 1,111_3$$
8. Write out the 3 times table in the 6-finger system. Does it ·remind you of any table in our usual arithmetic?
9. Write out the 1 times, 2 times, 3 times, and 4 times tables in the 5-finger system. Which table resembles the 9 times table in our usual system?
10. We write $7 \times 8 = 56$. In the 6-finger system this multiplication would be written $11_6 \times 12_6 = 132_6$. How would you write the multiplication $7 \times 7 = 49$ in the 6-finger system?
11. How would you write:

$4 \times 4 = 16$ in the 3-finger system?
$5 \times 5 = 25$ in the 4-finger system?
$6 \times 6 = 36$ in the 5-finger system?
$8 \times 8 = 64$ in the 7-finger system?
$9 \times 9 = 81$ in the 8-finger system?

Do you notice anything about your answers?

Divisibility

12. How would you write:

$7 \times 7 = 49$ in the 5-finger system?
$8 \times 8 = 64$ in the 6-finger system?
$9 \times 9 = 81$ in the 7-finger system?
$10 \times 10 = 100$ in the 8-finger system?

What do you notice?

13. In the same way, write:

$9 \times 10 = 90$ in the 7-finger system.
$10 \times 11 = 110$ in the 8-finger system.
$11 \times 12 = 132$ in the 9-finger system.

You may find it interesting to try to compose more examples like questions (10) to (13). An explanation of the coincidences observed here will be found later in this book.

14. Is the number $123,456_8$ even or odd? How do you test for evenness in the 8-finger system?

15. In the 7-finger system, how do you test (i) whether a number is even, (ii) whether it is divisible by 3?

Apply these tests to the number $1,111_7$.

CHAPTER THREE

An Unorthodox Point of Entry

IT is a defect of most algebra books that they begin by developing a lot of machinery, and it is a long time before the learner sees what he can do with all this machinery. For example, he may learn to simplify $5(x+3)-4(2-x)$ without seeing in just what circumstances he would feel a need to perform this calculation.

It is quite possible to use simultaneous equations as an introduction to algebra. Within a single lesson, pupils who previously did not know what x meant, can come, not merely to see what simultaneous equations are, but to have some competence in solving them. No rules need to be learnt; the work proceeds on a basis of common sense. The problems the pupils solve in such a first lesson will not be of any practical value. They will be in the nature of puzzles. Fortunately, nature has so arranged things that until the age of twelve years or so children are more interested in puzzles than in realistic problems. So the puzzle flavour of the work is, if anything, an advantage. The children get a sense of achievement, which they do not always get when they simplify $5(x+3)-4(2-x)$.

So we start with a puzzle. 'A man has 2 sons. The sons are twins; they are the same height. If we add the man's height to the height of 1 son, we get 10 feet. The total height of the man and the 2 sons is 14 feet. What are the heights of the man and his sons?'

Here we have a mass of words. Quite likely some boy on the back row has not taken in their full meaning. So we do not start on the job of solving the problem. We first try to make sure that

Figure 10

40

An Unorthodox Point of Entry

we can see what the problem means – for if we do not understand the question, we have no hope of finding the answer. The first question posed to the class is – how are we going to picture this?

It is not hard to picture the man and his 2 sons (Figure 10).

Next, we are told something about adding the man's height to the height of 1 son. How shall we draw a picture to show their heights being added? I have tried this question on all kinds of audiences, from young children to professors. They all produce the same answer – the son must stand on his father's head.

Figure 11

Figure 11 shows that their heights add up to 10 feet.

It is now easy to picture 'the total height of the man and the 2 sons'. The second son must climb up on to the top of the first

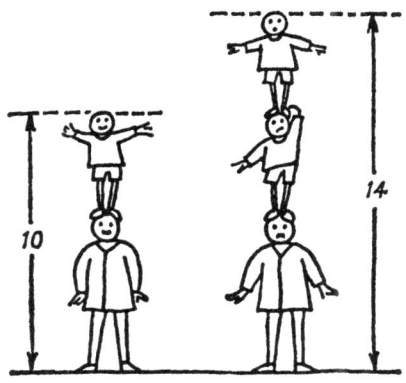

Figure 12

41

son's head. So in Figure 12 we have our problem stated in picture form.

Once this stage has been reached, it is quite unnecessary for the teacher to tell the class what to do. They will say, spontaneously, 'The sons must be 4 feet high.' Why? Well, when the second son climbed on, the height rose from 10 to 14 feet, so he must have brought the extra 4 feet. And, as the man and one son fill a space of 10 feet, the man must be 6 feet high.

We may get tired of drawing men and boys, and decide to use, instead of the complete picture, a simplified drawing as in Figure 13.

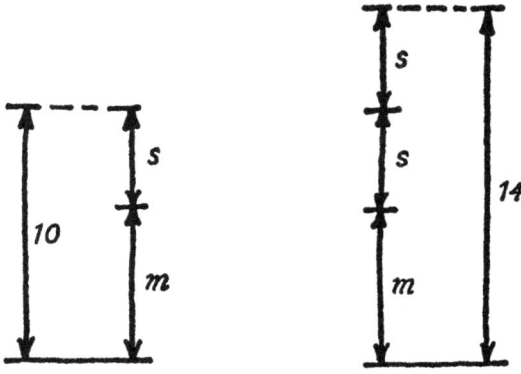

Figure 13

Where before we had a man, we now have an arrow marked m; where before we had a son, we now have an arrow marked s.

What does our diagram show? On the left, we see that m and s put together make 10. How shall we record this? The pupils will suggest that we write $m+s=10$.

On the right of the diagram we see that m and $2s$ make up to 14. It seems quite natural to record this as $m+2s=14$.

Thus we may write our problem as:

$$m+s=10 \qquad \text{(i)}$$
$$m+2s=14 \qquad \text{(ii)}$$

We now go back to our original picture. We think how we

42

An Unorthodox Point of Entry

solved the problem, and we follow exactly the same line of thought with the equations above.

With the picture we began by observing that the height rose from 10 to 14 feet when an extra son climbed on. In looking at equations (i) and (ii), we notice the same thing. There is an extra s in (ii). This must be responsible for the extra 4 in 14 as compared with 10.

So we write:

$$\therefore\ s=4$$

With the picture we argued, 'The man and a son fill 10 feet. We know the son is 4 feet high. The man must be 6 feet.'

We do the same now. We have found out that s stands for 4. So we go back to equation (i) and use this information there.

$$m+s=10$$
$$\therefore\ m+4=10$$
$$\therefore\ m\ \ \ \ =6$$

The exact meaning of m and s here should be noted. They do not stand for 'the man' and 'a son'. It is not correct to say 'the man equals 6' (though a child might say this in an unsuccessful attempt to express a correct idea). Nor does m even stand for 'the height of the man'. The man's height is 6 *feet*. The letters m and s do not stand for human beings, nor for lengths: they stand for *numbers*; m for the number of feet in the man's height, s for the number of feet in a son's height.

In certain engineering work, letters are used to represent lengths and weights. In beginning algebra this can lead to confusion. If a student thinks of x as meaning so many feet, he will have to think of x^2 in terms of square feet, x^3 as cubic feet, and I do not know what he will do about x^4. He will be puzzled by an ordinary quadratic equation such as $x^2+5x=6$, for here (he will think) we are adding square feet to feet, and getting a plain number as the sum. All these complications and difficulties are avoided if, *at the stage of beginning algebra*, letters are used to indicate numbers only. Later on in mathematics, letters may be used to indicate anything we like.*

* See *Prelude to Mathematics*, W. W. Sawyer, Chapter 7.

Vision in Elementary Mathematics

It is not enough for a child to hear an idea explained, and to understand it. If that is all we do, the idea may leave no trace in the child's memory at all. Children need to spend some time with the idea, to play around with it, to use it themselves.

Textbooks for this reason contain masses of exercises for the pupils to work. Sometimes these are dull. Sometimes they go too quickly to difficult problems. One way to enliven the teaching of simultaneous equations, and to keep the work at the children's own level, is to get the pupils to invent the problems. We say, 'Think of a height for a man, and a height for his sons. Do not tell me these, but tell me, first, what you get when you add the man's height to the son's, and second, what you get when you add the man's height to twice the son's height'. If the answers to these questions were, say, 7 feet and 9 feet, we should write:

$$m+ \ s=7$$
$$m+2s=9$$

Very likely, the children can complete the solution of these equations as they are written. If so, well and good. If not, we go back to drawing pictures, and think the whole thing through again. We shall find $m=5$ and $s=2$.

Giants and dwarfs are perfectly in order. The children may produce men 200 feet high, and sons $\frac{1}{2}$ foot high. Variety will come naturally into the arithmetic; children will experiment by introducing large numbers and fractions.

Some writers of textbooks get into great difficulties, because they try to start with a definition. They will say perhaps, 'In algebra, letters such as x, m, s stand for unknown numbers.' Then someone else will criticize this definition. In our example above, the child who made up the problem knew what m and s stood for; the other children did not know until they had solved the problem. Was m a known or an unknown number? Known to whom?

Almost any short sentence you can say, attempting to define what we mean by x, will be wrong. Not only that; it will be an ineffective approach to teaching. We do not give a baby a scientific definition of 'dog' – but babies know what dogs are. Much the best way to introduce children to algebra is to show them algebra in action.

An Unorthodox Point of Entry

The procedure used above conveys several important ideas that would be very difficult to explain in words. In the first problem $m=6$; in the second problem $m=5$. So m means different things at different times. Yet the meaning of m has to be held fixed in certain circumstances. If Tom Smith proposes a problem about a man and his sons, so long as we are considering that problem, m must signify the height in feet of the man imagined by Tom Smith. So long as we are working at this problem, each time we write m it has the same meaning. When we have found out what Tom Smith was thinking, we can go on to a problem invented by Henry Brown, and begin, 'Let m feet stand for the height of the man Henry Brown is thinking of '.

VARIATIONS ON A THEME

We now continue with our exploration of simultaneous equations. We gradually bring in new variations.

So far, all our problems have been concerned with $m+s$ and $m+2s$. But we can solve other problems, for example:

$$m+\ \ s=7$$
$$m+3s=10$$

Here a man evidently has 3 sons all the same height. We could, if necessary, draw pictures to illustrate these equations. 1 son standing on the man's head achieves a height of 7 feet; 3 sons on the man's head form a tower 10 feet high. By adding 2 sons, we have increased the height by 3 feet. We may write:

$$2s=3$$

If 2 sons occupy 3 feet, each son must be $1\frac{1}{2}$ feet.

$$s=1\frac{1}{2}$$

The man must be $5\frac{1}{2}$ feet high.

$$m=5\frac{1}{2}$$

Again, the pupils are invited to compose variations on this theme.

Vision in Elementary Mathematics

Nor need there be only 1 man. We imagine a crowd of men, all the same height, and an unlimited supply of sons, who are also a standard height. We can then consider equations such as:

$$3m+2s=24$$
$$3m+4s=30$$

Again, if it seems necessary, we can draw pictures. 3 men and 2 sons form one tower; 3 men and 4 sons the other. The towers differ in height by 6 feet, and this difference is caused by 2 more sons climbing on.

$$2s=6$$
$$s=3$$

Now we go back to the first equation, and use the fact that s stands for 3.

$$3m+2s=24$$
$$3m+\ 6=24$$
$$3m\quad\ \ =18$$
$$m\quad\ \ =6$$

If these steps are not clear, they can all be worked out from the pictures. In the first tower we see 3 men and 2 sons forming a height of 24 feet. But we have discovered that the sons are 3 feet high. So 2 sons fill 6 feet of the tower. The 3 men must fill the other 18 feet. So each man must be 6 feet. The equations written above are simply a short way of recording this argument.

AN IMPORTANT DEVICE

In all the examples considered so far, the second tower was built from the first tower by allowing more sons to climb on. By considering the extra room needed by the extra sons, we rapidly found the height of a son.

But not all problems are like this. Suppose we are told:

$$m+\ s=10 \qquad\qquad \text{(i)}$$
$$3m+5s=36 \qquad\qquad \text{(ii)}$$

Here the second tower contains 2 men and 4 sons more than the first. We cannot tell how much of the extra 26 feet is due to the

46

men and how much to the sons. So we cannot proceed as we did before.

Fortunately, there is a simple trick that will help us out. The second tower (corresponding to equation (ii)) contains 3 men and some sons. We would have been happy if the first tower had also contained 3 men and some sons, for then we would have been able to tell the effect of the sons. As the first tower does not contain 3 men, we build one that does.

This is not difficult. If a certain collection of people take up a certain amount of room, twice that collection will require twice as much room, 3 times that collection 3 times as much room, and so on.

We know that 1 man and 1 son take up 10 feet. By doubling, we see that 2 men and 2 sons fill 20 feet. In fact, we can make a long list of statements as in Figure 14.

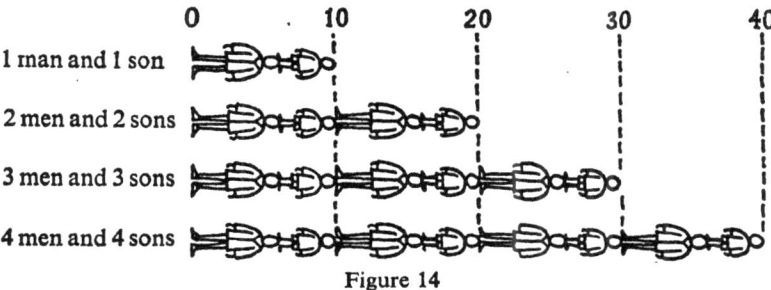

Figure 14

(This illustration has been put on its side purely for reasons of paper economy.)

1 man and 1 son fill 10 feet
2 men and 2 sons fill 20 feet
* 3 men and 3 sons fill 30 feet *
4 men and 4 sons fill 40 feet
and so on.

In this list, the stars indicate a statement that is particularly useful to us, for it is about 3 men and some sons – just what we wanted. In the language of algebra it reads $3m + 3s = 30$. We call this equation (iii).

Vision in Elementary Mathematics

If we now write equation (ii) and equation (iii) together, we recognize a type of problem that we know how to solve. Here are the equations.

$$3m+3s=30 \tag{iii}$$
$$3m+5s=36 \tag{ii}$$

These differ only in the number of sons involved. The pictures of these 2 equations show us that 2 more sons require 6 more feet.

$$2s=6$$

So $s=3$. Now we may as well go right back to equation (i). A man and a son fill 10 feet. So the man must be 7 feet high, $m=7$.

To use this method correctly, a pupil must understand two things.

First, he must see that the statement 'a man and a son fill 10 feet' allows us to make many other statements. We can be sure that 3 men and 3 sons will fill 30 feet; 17 men and 17 sons will fill 170 feet; 100 men and 100 sons will fill 1,000 feet. These and many more we know for certain, the men and boys being standard sizes.

All of these statements are true. The second thing the pupil must be able to do is to pick out the statement that will be most useful as a clue to the solution. He must look carefully at equation (ii), which states $3m+5s=36$. This equation starts by talking about '3 men'. If we choose out of all the statements above the statement which begins with '3 men', we shall have something that joins on well to equation (ii). The pupil will probably see the point after a few examples have been demonstrated by the teacher.

For instance, if we are given the equations:

$$m+3s=11 \tag{i}$$
$$4m+7s=34 \tag{ii}$$

the second equation begins with '4 men and . . .'. We want to use equation (i) so that it also gives a statement beginning '4 men and . . .'. This is not hard to find, for equation (i) tells us that 1 man and 3 sons fill 11 feet. It follows that 4 men and 12 sons fill 44 feet. This is the statement we want for equation (iii). We write it down, together with equation (ii).

48

An Unorthodox Point of Entry

$$4m+12s=44 \qquad \text{(iii)}$$
$$4m+ 7s=34 \qquad \text{(ii)}$$

If we picture these, we see that there are 5 extra boys who fill the extra 10 feet. So we write:

$$5s=10$$
$$s=2$$

Now if we go back to equation (i), this shows us $m=5$.

The pupils themselves should, of course, work several exercises to fix the ideas in their memory. As before, the pupils can help to make up the exercises. At first, the teacher may have to give instructions in full. Some pupil is asked to choose heights for a man and a son. He keeps these secret, but the teacher asks him, 'How many feet would a man and 3 sons fill? How many feet would 4 men and 7 sons fill?' The rest of the class now have to work out the heights of a man and a son.

As the pupils become familiar with the symbols m and s, the instructions become briefer. 'Choose numbers for m and s. What do you get for $m+3s$? What do you get for $4m+7s$?' From the answers, the class deduce the numbers chosen for m and s.

No one should ever hand in an incorrect solution to such a problem. For we can always test our answers, and see whether they fit the conditions imposed by the problem.

A person learning by himself can make up his own problems. He may compose, say, a dozen questions, and keep a record of the numbers he chose. A day or two later, when he has forgotten the numbers, he can work through the questions. As soon as he has worked each question, he should test it, and then immediately look at his record of the solution, to make sure that he is working correctly.

A few exercises are given here, but a learner should work far more than these. Answers are given at the back of the book.

Exercises

Find m and s in each of the following problems:

1. $m+ s=10$	**2.** $m+ s=8$	**3.** $m+ s=7$
$m+2s=11$	$m+5s=20$	$2m+3s=16$

49

4. $m+\ s=8$ **5.** $m+2s=10$
   $2m+\ 4s=21$ $2m+5s=22$

6. $m+\ 3s=11$ **7.** $m+\ 4s=14$
   $10m+33s=116$ $100m+399s=1,398$

MORE FORMAL WORK

In the beginnings of arithmetic and algebra, the main purpose is not to get the pupil making calculations. The main purpose is to get him into the habit of thinking, and to show him that he can think the problems out for himself.

That has been the aim of this chapter so far. The examples have been made as simple as possible.

At some stage, the pupils will have to acquire technical skill. They will not only have to understand the ideas used in solving simultaneous equations; they will have to be able to solve any system of equations they may meet.

This technical part should be taught at a stage when the *ideas* have settled down in the minds of the pupils, and have been firmly established.

It will be convenient to find a label for an operation that we have used and shall use frequently.

On page 48 we used the argument:

$$\text{If} \quad m+\ 3s=11 \qquad \text{(i)}$$
$$\text{then} \quad 4m+12s=44 \qquad \text{(iii)}$$

We justified this by saying that if we had 4 times as many men and 4 times as many sons, they would need 4 times as much room. So everything has been multiplied by 4 – the number of men, the number of sons, the number of feet filled. If you look at equation (iii) above, the numbers in it are just 4 times the numbers in equation (i).

It is convenient to indicate this by writing:

$$4 \times \text{(i)}$$

50

An Unorthodox Point of Entry

If I said to you, 'Write the equation $10 \times$ (i)', this would mean, 'Take equation (i) and multiply everything in it by 10.' This would give you:

$$10m + 30s = 110 \qquad \text{(iv)}$$

If the equation (i) was a true statement, equation (iv) also would be a true statement. For with 10 times as many men and 10 times as many sons, you could fill 10 times as many feet.

In all the problems we considered earlier, equation (i) was a statement about *one* man and several sons. This made the problem easy to solve. For example, to solve the problem:

$$m + 3s = 13 \qquad \text{(i)}$$
$$10m + 28s = 124 \qquad \text{(ii)}$$

we form the equation $10 \times$ (i). This gives:

$$10m + 30s = 130 \qquad \text{(iii)}$$

Compare (ii) and (iii):

$$2s = 6$$
$$s = 3$$

Go back to (i): $\qquad m = 4$

But now suppose we meet the problem:

$$2m + 5s = 27 \qquad \text{(i)}$$
$$3m + 4s = 30 \qquad \text{(ii)}$$

It would be possible to solve this by forming the equation $1\frac{1}{2} \times$ (i) and comparing it with (ii). However, many people prefer to avoid fractions, and indeed all the questions treated in this chapter could be given as puzzles in arithmetic to children who had not yet learnt fractions.

We can find a 'fraction-free' treatment of the problem. We have to form two new equations, $3 \times$ (i) and $2 \times$ (ii).

$$3 \times \text{(i) gives } 6m + 15s = 81 \qquad \text{(iii)}$$
$$2 \times \text{(ii) gives } 6m + 8s = 60 \qquad \text{(iv)}$$

Equations (iii) and (iv) refer to the same number of men; by comparing them we can get a statement involving sons only. In fact it is:

$$7s = 21 \qquad \text{(v)}$$
$$s = 3 \qquad \text{(vi)}$$
Going back to (i), $\qquad m = 6 \qquad$ (vii)

It should be noted that this is a logical process. Pupils ask 'Am I allowed to do this?' as if we were playing a game with certain rules. A pupil is allowed to write anything that is true, and not allowed to write anything untrue! These are the only rules of mathematics.

The person who poses the problem is responsible for statements (i) and (ii). He says to us, 'I have thought how high a man might be, and how high a son. With the heights I have chosen, 2 men and 5 sons fill 27 feet; 3 men and 4 sons fill 30 feet.' We hope he has not made a mistake in his arithmetic. We accept his statements as true.

Now we argue: if 2 men and 5 sons fill 27 feet, *it must be* that 6 men and 15 sons fill 81 feet. Equation (iii) records this conclusion.

In the same way, equation (ii) (which we have accepted as true) says that 3 men and 4 sons fill 30 feet. We argue, *it must be* that 6 men and 8 sons fill 60 feet. Equation (iv) records this.

Comparing (iii) and (iv), we see that the tower for equation (iii) contains 7 more sons than the tower for equation (iv). *It must be* these extra 7 sons that cause the extra 21 feet in height. Equation (v) expresses this.

So *it must be* that each son is 3 feet high (equation (vi)). And going back to the first statement made (equation (i)) we see that a man *must be* 6 feet high.

The whole argument then shows that if a person has thought of a height for a man and a height for a son, and has made true statements (i) and (ii), then he cannot have been thinking of any heights other than 6 feet for the man and 3 feet for the son. And we check that these actually do solve the problem.

EQUATIONS WITH FRACTIONS

Another complication that can arise is the presence of fractions. The proposer of the problem might begin by saying, 'Half the height of the man added to one third of the height of the son gives 4 feet.'

The corresponding equation would be:

$$\tfrac{1}{2}m + \tfrac{1}{3}s = 4 \tag{i}$$

A pupil coming to a collection of such problems in an algebra book feels this is going to be much worse -- the fractions look much harder than the whole numbers!

Actually, it is easy to get rid of the fractions. We apply to equation (i) the principle that 6 times as many people need 6 times as much room; that is, we form the equation $6 \times$ (i). As 6 half-men make 3 men and 6 thirds of a son make 2 sons, we get:

$$3m + 2s = 24$$

There are no fractions here; we have replaced equation (i) by an equation containing whole numbers.

The second statement in the problem might give us the equation:

$$\tfrac{1}{6}m + \tfrac{1}{4}s = 1\tfrac{3}{4} \tag{ii}$$

We might then form $12 \times$ (ii), which is:

$$2m + 3s = 21$$

Thus the original problem:

$$\tfrac{1}{2}m + \tfrac{1}{3}s = 4 \tag{i}$$
$$\tfrac{1}{6}m + \tfrac{1}{4}s = 1\tfrac{3}{4} \tag{ii}$$

has been replaced by:

$$3m + 2s = 24 \tag{iii}$$
$$2m + 3s = 21 \tag{iv}$$

The work would now be completed in the way described in the previous section.

Of course a certain amount of judgement was needed when we decided to form $6 \times$ (i) and $12 \times$ (ii). These numbers, 6 and 12, are the smallest numbers that will get rid of the fractions.

Needless to say, this type of problem should only be attempted by pupils who already have a good understanding of fractions.

EQUATIONS INVOLVING SUBTRACTIONS

A person inventing a problem, we will suppose, has decided to have the man 6 feet high and the sons 4 feet. This person might then put the following puzzle to us. 'A man and 2 sons fill 14 feet. If you subtract the height of 1 son from the man's height, you get 2 feet. How high are the man and the sons?'
His statements would be written:

$$m+2s=14 \qquad\qquad \text{(i)}$$
$$m- s=2 \qquad\qquad \text{(ii)}$$

(Notice, in passing, that subtracting the son's height from the man's gives $m-s$, NOT $s-m$. You can see this by considering that $m=6$ and $s=4$; we want $6-4=2$, *not* $4-6$.)
How are we going to picture $m-s$? Figure 15 shows one way.

Figure 15

The son is hanging down, so that his height is *taken away* from the man's.
The full picture of equation (i) and (ii) would therefore be as in Figure 16.
From this picture we can draw a useful conclusion. We notice that the heavy line at the height of 14 feet is separated from the heavy line at the level of 2 feet by a space into which 3 sons can be fitted. This is, 3 sons fill 12 feet, so each son is 4 feet.
The higher heavy line is at height $m+2s$; the lower heavy line is at height $m-s$. The argument we have just used shows that $m+2s$ is bigger than $m-s$ by the amount $3s$.
We might save ourselves the trouble of drawing men and sons.

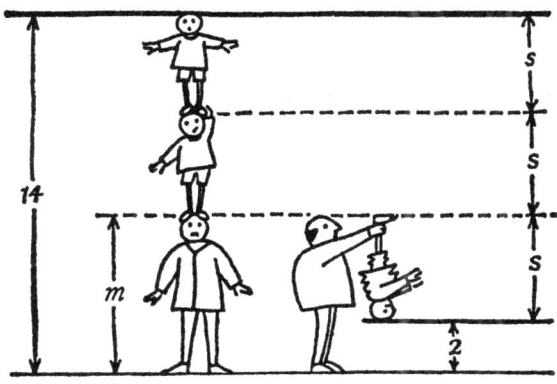

Figure 16

Suppose we use a black arrow to represent a man and a white arrow for a son.

We can then picture $m+2s$ and $m-s$ as in Figure 17.

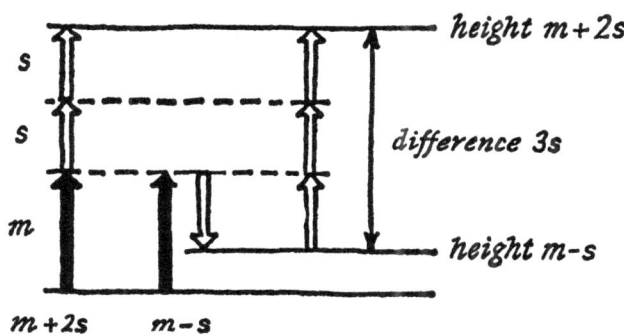

Figure 17

If we are told that $m+2s$ is 14 and $m-s$ is 2, then the difference $3s$ must be the difference between 14 and 2, that is, 12.

So the solution of the problem would look like this:

$$m+2s=14 \qquad\qquad \text{(i)}$$
$$m-\ s=2 \qquad\qquad \text{(ii)}$$

Comparing (i) and (ii) $\qquad 3s=12 \qquad\qquad \text{(iii)}$

Vision in Elementary Mathematics

$$s=4 \qquad \text{(iv)}$$

Going back to (i) $\qquad m=6 \qquad$ (v)

In writing equation (iii) we are helped by the diagram above.

Exercises

1. In Figure 18, what are the two heights, and what is the difference between them?

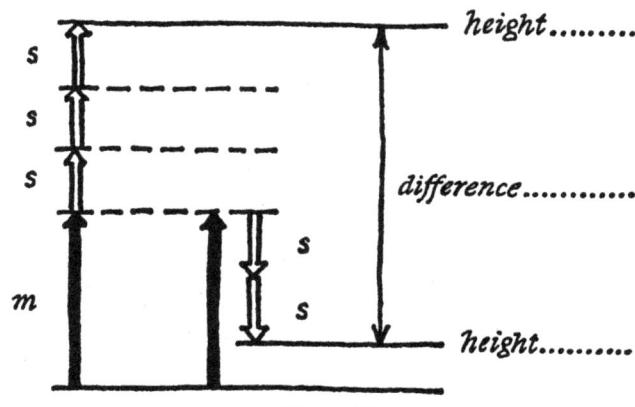

Figure 18

2. How much larger is $m+s$ than $m-s$? (Draw diagram.)
3. How much larger is $m+2s$ than $m-3s$? (Draw diagram.)
4. How much larger is $m+6s$ than $m-4s$?
5. How much larger is $m+17s$ than $m-11s$? (You may not wish to draw all 17 white arrows pointing up and all 11 pointing down. You can indicate on a simplified diagram that they are supposed to be there.)
6. In Figure 19, what are the two heights? What is the difference between them?
7. Which is larger, $m-s$ or $m-3s$? What is the difference between them?*

* For the benefit of any mathematician who may chance to read this, I would point out that the idea of a negative value for s is, at this stage, entirely outside the children's thinking. It would only confuse them if we stated an extra condition, $s > 0$.

56

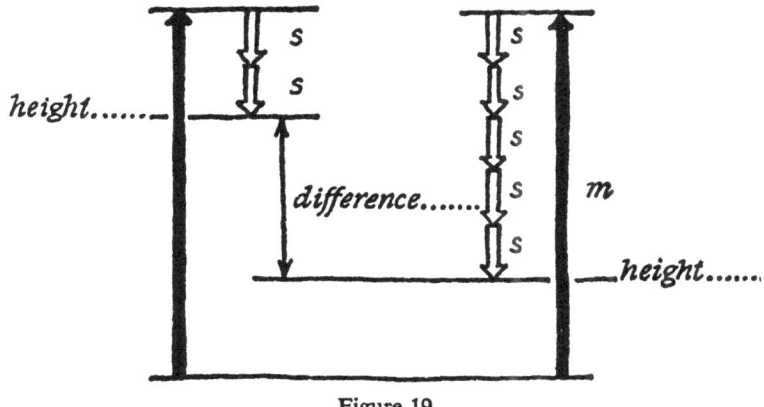

Figure 19

One final point remains to be mentioned. Suppose we have the equation:

$$m-s=10 \qquad \text{(i)}$$

We may need to form the equation $2 \times$ (i). It can be shown that the correct result is:

$$2m-2s=20$$

I mean by this, that if numbers m and s have been chosen such that $m-s=10$, then *it must be* that $2m-2s=20$. You can test this by choosing particular numbers. If $m=11$ and $s=1$, $m-s=10$, and $2m-2s=22-2=20$. You will not be able to find numbers m and s that make (i) true but make $2m-2s$ not equal to 20. Why this is so will be discussed in Chapter 10.

In the same way, if equation (i) were:

$$2m-3s=4$$

and we wanted to form $5 \times$ (i), the result (as you might well guess) would be:

$$10m-15s=20$$

A simple application of these ideas would be the following problem.

$$m-s=4 \qquad \text{(i)}$$
$$2m+s=14 \qquad \text{(ii)}$$

Vision in Elementary Mathematics

$$2 \times \text{(i) gives } 2m - 2s = 8 \qquad \text{(iii)}$$

Comparing (ii) and (iii) $\qquad 3s = 6 \quad$ (Draw diagram!)

$$s = 2$$

From (i) $\qquad m = 6$

TEACHING DIFFICULTIES

One of the things you find in teaching is that you can give a very clear explanation, and still the pupils – or some of them – will not understand. A poor teacher simply repeats the original explanation, like a robot. A good teacher tries to find out exactly where the pupil failed to understand. This is the essence of teaching – to try to guess what is happening in the pupil's mind, and to lead the pupil's thinking to the next stage. If we imagine that a pupil understands something, which in fact he does not, we are like a man trying to lay bricks on a foundation of air.

I am sure that in this book there will be places where I have tried to develop an argument as clearly and carefully as possible, and yet some reader will fail to understand. I have tried to foresee every difficulty, but some reader will begin with a viewpoint different from mine, and will meet obstacles that I never imagined.

A teacher has much the same difficulty when he has to deal with a large class. There is not enough time for talking to individual pupils, so the teacher works in the dark. Some pupil may have misunderstood some small point, and this difficulty is holding up his entire progress. Five minutes discussion might clear it up, if only the teacher knew what the difficulty was.

Suppose a pupil has been taught along the lines explained in this chapter. He is given questions on this work, and his answers are wrong – not merely because of errors in arithmetic, but wrong in a way that shows failure to understand. How shall we deal with this? Each part of this chapter has assembled certain material – statements, questions, problems, pictures – in the hope that these would lead the pupil to a certain idea. If at the end of any part, the pupil has not formed this idea, the explanation has failed. The pupil has heard the statements, seen the pictures, made some attempt to work the exercises, but everything has remained meaningless; the pupil has not seen what I was trying to convey.

An Unorthodox Point of Entry

If I am to correct this, I must find out at which stage I failed to establish communication.

We began with pictures to illustrate puzzles. We first test whether the child has grasped this idea. We ask him to illustrate statements such as those in our first problem, 'the heights of a man and his son add up to 10 feet'. Most children do this easily enough.*

The next principle is that of comparison. One tower contains a man and 2 sons, and is 14 feet high. Another tower contains a man and 1 son, and is 10 feet high. We make sure that the child can see that the extra 4 feet is due to the extra son. We test him by similar examples.

So far the work has been concrete. We have been dealing with actual things – men, sons, and the space they occupy. Difficulties are rare at this stage.

The next step is a big one, when we pass to equations such as $m+s=10$. A child can look at $m+s=10$ and simply see meaningless marks on paper. We have to check carefully that the child can write the equation corresponding to any picture, and draw the picture corresponding to any equation. This must be done often enough for pictures and equations to become closely associated in the child's mind. Any time there is difficulty in reasoning about an equation, the picture of it should be drawn.

And here we meet one of the difficulties involved in abstraction. We draw a picture showing a man with his son standing on the man's head, to represent $m+s$. It may happen that we draw the man twice as tall as the son. The child may start to reason from this fact; if $m+s=10$, and the child thinks the man in the picture *looks* twice as tall as his son, he will argue (if his arithmetic is good enough) that the man must be $6\frac{2}{3}$ feet high and the son $3\frac{1}{3}$. We have to point out that the picture is only meant to show that the *total height* is 10 feet. It was based only on this information, and we have no way, yet, of deciding which of the pictures in Figure 20 is most true to life.

We can discuss the various possibilities for $m+s=10$. It may be that the man is 9 feet and the son 1 foot high. The man may

---

* I am not sure how it would work with a person who thought in terms of sound rather than sight or touch.

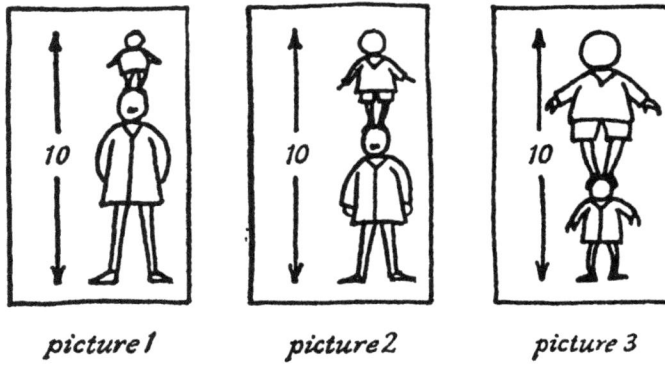

picture 1   *picture 2*   *picture 3*

Figure 20

be 7 and the son 3. In a puzzle we even have to consider that the man might be 2 and the son 8 feet high. It is only when we come to the second statement in the puzzle, $m+2s=14$, that we find most of our guesses fail to fit.

Children, of course, often guess their way to the solution. Do not say, 'It is worthless unless you do it by the proper method.' Mathematics is meant to encourage thinking, and guessing is one way of thinking. In fact, guessing helps to convey to a child the meaning of the puzzle. If a child guesses $m=7$ and $s=3$, we ask him to check whether this fits both equations. He finds, yes, it does fit $m+s=10$, but no, it does not fit $m+2s=14$. A few guesses help to drive home the fact that a solution must fit *all* the statements of the puzzle.

Mathematicians regularly guess solutions of their problems. Systematic methods of solution are needed when problems are too complicated for the answer to be guessed. One can wean children away from pure guesswork by posing problems in which the numbers are too large for the solution to be guessed. The child then sees why some system for solving is necessary.

Suppose then a child has learnt to associate equations and pictures, and to use the pictures properly – that is to say, he does not argue from accidental details of the picture. We make sure that he can carry over to written equations (illustrated if necessary by pictures) the simple arguments that were used to solve the

An Unorthodox Point of Entry

first problem in this chapter. We should not be surprised if he finds difficulty in the section headed 'An Important Device'. Some children pick up the ideas of this section very quickly. Others require much time to think about them. There are two ideas involved, and both are liable to cause difficulty. First, a child may not see that a statement such as $m+s=10$ necessarily carries with it the truth of other statements such as $2m+2s=20$ and $7m+7s=70$. Second, even after having understood this first point, a child may not know which of all these possible statements is the one that will help solve a particular problem. And indeed, it is not surprising that an understanding of this second point should often come slowly, for it is an example of an advanced type of thinking – what one might call strategic planning. The child has to say to himself, 'Here is this equation. I could multiply it by 2, or by 3, or by 4, or by many other numbers. Which of all these will give me a result that combines nicely with the other information I have about this puzzle?' To answer such a question requires a fairly wide mental horizon. A child cannot be expected to make a wise choice about which operation to perform while he is still puzzled about the details of the operations themselves. Nor can this wise choice be expected until the child has solved enough problems successfully for him to see the process of solution as a whole. It is, therefore, quite in order for a teacher to set problems (at any rate for the average pupil) with a hint, 'multiply equation (i) by 4'.

The power of strategic planning is the most important element in the solving of problems, and it is the one that is least taught. Rote teaching bypasses it completely. The pupil is told what to do, so he never learns to weigh one method of attack against another. A pupil can only learn good judgement by seeing the effects of bad judgement. It is instructive to work one problem by five or six different methods, some of which may lead to blind alleys and have to be abandoned, some of which may give the correct solution but only after laborious wanderings, and some of which give the solution with a minimum of effort.

There should be plenty of discussion of what method to use, and different methods should be tried out. This is emphasized, because a teacher may feel that it is a waste of time to explore

61

approaches that lead to no solution. It is by no means wasted time. We only appreciate and understand a good method if we have seen the consequences of using bad methods.

So much for strategic planning; we return to the first and smaller difficulty, that of the child who sees no connexion between $m+s=10$ and $2m+2s=20$.

We have already made some effort to treat this pictorially. But, of course, in teaching it is not sufficient to show a picture once. It has to be pulled out many times, and in the end its message sinks into the learner's mind. A picture is, in fact, a most valuable way of reminding someone of a sustained chain of thought. You show the picture and ask, 'Do you remember the discussion we had about this?'

Our earlier picture pointed out that if a man and his son fitted into a space of 10 feet, by simply repeating the drawing you could see a way of filling 20 feet.

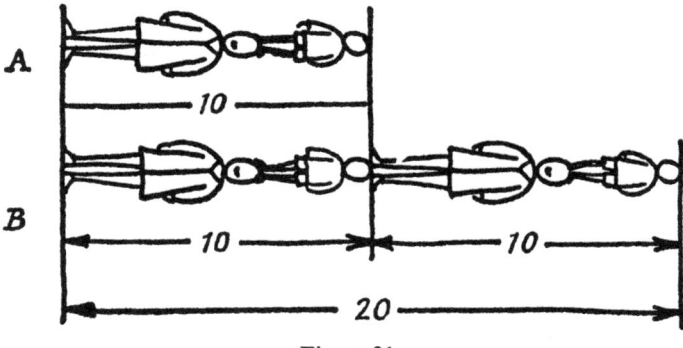

Figure 21

Figure 21 shows that, if 10 feet holds a man and a son, 20 feet will hold 2 men and 2 sons. But perhaps it does not show this as clearly as it might. If we were asked to illustrate '2 men and 2 sons', the natural thing to draw would be as in Figure 22.

Figure 22

62

An Unorthodox Point of Entry

It is not immediately obvious that this collection just fills 20 feet. Collection C contains 2 men and 2 sons, as does collection B, but the arrangement is different. C suggests $2m+2s$ where B suggests $m+s+m+s$. So, while it is immediately evident to the eye that B takes twice as much room as A, it is not immediately evident that C does. It can be appreciated by the mind, if we recognize that the change in the order of the men and the sons, as we go from B to C, does not affect the total space occupied.

A demonstration with movable objects, such as bricks or rods, might help children to appreciate the connexion between A, B, and C.

At any rate, the pictorial approach should help to bring a child some distance towards feeling (and remembering) that the statement, 'If $m+s=10$, then $2m+2s=20$' is a reasonable one.

We supplement the pictures by experiments in arithmetic. Different pupils are asked to think of 2 numbers that add up to 10. The first number any pupil has thought of we call m for short, the second s. So for each pupil, $m+s=10$. We then ask round the class, 'What did you get for $2m+s$? What for $2m+2s$? What for $2m+3s$?' The answers are tabulated like this:

	Jack	Mary	Anne	Bill
$2m+\ s$	18	13	19	15
$2m+2s$	20	20	20	20
$2m+3s$	22	27	21	25

It now stands out very clearly that 20 occurs all the way across the middle row. The class may try to find 2 numbers adding to 10 that make $2m+2s$ different from 20. They will fail to find an exception (except any due to faulty arithmetic).

Further, in looking at the table above, the children will notice how each column goes up by equal steps. Jack has 18, 20, 22 rising by 2, Mary has 13, 20, 27 rising by 7, and so on. And the other pupils will often spontaneously say, 'Jack was thinking of 8 and 2,' and in the same way they will discover what numbers the others chose. If they do this by guesswork, they may notice that the second number, s, is the same as the step by which the numbers rise; for example, Jack had $s=2$ and his numbers 18, 20, 22 rise by steps of 2. And in this way they will be led back to

63

Vision in Elementary Mathematics

the principle of comparison, for the expressions $2m+s$, $2m+2s$, and $2m+3s$ rise by steps of s.

Children obtain a very firm grasp of algebra if they are given time to experiment, observe, and argue. Some of the most interesting arguments arise from errors. For example, in one class, children were asked to think of numbers for m and s – the heights in feet of a man and his son – and to give certain information. One girl said, 'With my numbers $m+s=10$ and $2m+3s=17$.' This was hotly questioned. One pupil pointed out that if $m+s$ was 10, then $2m+2s$ must be 20, and $2m+3s$ must be more than 20. So $2m+3s=17$ must be wrong. In fact, the girl had made an error in her mental arithmetic.

It appears from the table above (about Jack, Mary, Anne, and Bill) that, if we know $m+s=10$ we are able to say what $2m+2s$ is, but we cannot predict the values of $2m+s$ and $2m+3s$. One way to test understanding of this work is to give a pupil a list of expressions, such as $3m+s$, $3m+2s$, $3m+3s$, $3m+4s$, $4m+s$, $4m+2s$, etc., and ask him to pick out those whose values are fixed when we know $m+s=10$, and to say what these values are. Sometimes you will find a child giving values for all the expressions. This may mean he has thought, 'We would have $m+s=10$ if m was 6 and s was 4', and he has worked on the assumption that $m=6$ and $s=4$. Naturally, if we knew the man was 6 feet and the son 4 feet we could work out the height of any combination of men and sons. The problem we have put to the child is what conclusions we can draw if we only know the total height of the man and the son. If the child makes a table of the Jack – Mary – Anne – Bill type, he will find that a variety of numbers appear in most rows; we get a fixed number only for $3m+3s$, $4m+4s$, $5m+5s$, and so on.

It is just as important to realize that the value of $3m+2s$ cannot be deduced from that of $m+s$, as it is to know that $4m+4s$ can.

CHAPTER FOUR

Tricks, Bags, and Machines

IN the last chapter it was pointed out that simultaneous equations could be used as a starting point in algebra, but, of course, it is not compulsory to start with them. The present chapter discusses some other devices that can be used to introduce children to algebra. Any such device should satisfy certain conditions.

First, any child should be able to use it with success, so that it builds confidence. It is very bad if a first lesson leaves the pupil with the feeling that the new subject is mysterious, difficult, impossible.

Second, the device should call out the child's own powers of discovery and reasoning. The child should not feel that he has been told what to do; rather he should feel that his attention has been called to a certain problem, which he has solved by his own native wit.

Third, the result must be in some way intriguing, fascinating, remarkable. Children love mental excitement. Often a teacher has to struggle against inattention in a class. Some children look out of the window; others pass notes, or make drawings, or torment each other. But there are rare moments when the whole class concentrates on the business in hand. They are no longer trying to please the teacher or to keep out of trouble; they want to know for themselves. Almost invariably the topic that produces such attention has strong intellectual interest: it is beautiful, or surprising, or puzzling. Our teaching device should have as much as possible of this kind of appeal.

These conditions to some extent pull in opposite directions. The first two demand that the theme be simple; the third, that it be unexpected. But such a combination is not impossible. The same problem arises in teaching art or English composition. Children have little technical skill in the handling of words or materials, yet they wish to produce something creditable. They are grateful if we can show them how to produce a striking result by simple means.

Vision in Elementary Mathematics

There are certain tricks, such as card tricks not involving sleight of hand, which arouse curiosity. A surprising result is obtained. How is it done? Often the explanation is quite simple. In *Mathematician's Delight*, a pictorial interpretation of a well-known 'Think of a Number' trick was given. It may be worth while to repeat this explanation and to develop it a little further, as I have found it useful for introducing children to algebra. With young children, the trick itself is a surprise. Older children probably know the trick; the emphasis has to be on the questions, 'Why does the trick work?' 'How are tricks like this invented?'

The trick itself runs – think of a number, add 3, double the result, take away 4, divide by 2, take away the number you first thought of. We then announce, 'Your answer is 1', and the person is supposed to wonder how we knew this without knowing what number had been thought of. Often with children, of course, a mistake is made in the working, and the answer obtained is not 1. However, on going through the calculation again, the mistake will be found. Such tricks, incidentally, help to produce arithmetical accuracy – because a child who makes a mistake loses the surprise. In most school classes, there is a working majority for the answer, 1, enough to make children who get other answers uneasy. Eventually, the errors are sorted out, and the question arises, 'Why does everybody get the same answer?'

We now suggest to the children a way of visualizing what is happening. We suppose we have a good supply of stones (or marbles, or sticks, or what you will). When a person thinks of a number, he puts that number of stones in a bag. Of course, he is careful not to let us see how many he puts in. This bag: ⊘ helps us to picture 'the number thought of'. The next instruction says, 'Add 3.' So we put 3 stones beside the bag. We now have the picture: ⊘ ooo of a bag and 3 stones, which represents the number thought of with 3 added. This should be sufficient to convey the idea: from now on we ask the children to tell us what pictures illustrate the rest of the calculation.

The trick continues, 'Double the result.' The children may want to argue about this. One child may suggest that doubling a bag and 3 stones gives 2 bags and 3 stones; another may favour a bag and 6 stones. Both answers are suggested by the sound of the words rather than by real thinking. To picture something doubled, we draw that thing, and then draw the same again. If we do that with a bag and 3 stones, we get 2 bags and 6 stones.

The children find the instruction 'take away 4' easy to follow.

It gives 2 bags and 2 stones.

Next 'divide by 2'. We share the above 2 bags and 2 stones equitably between two children. Each child gets a bag and a stone:

$$\mathcal{S}_o$$

Finally, 'take away the number first thought of'. Where is the number first thought of? It is in the bag. So we erase the bag. Our pictures now show a single stone: **o**. It no longer matters what number of stones the bag contained. Whatever number was originally thought of, the final answer will always be 1. Children usually give an amused laugh when this conclusion is reached.

I have never met any normal child or adult who was unable to follow this demonstration. Yet this work is already algebra.

If arithmetic deals with definite results such as $2+3=5$, while algebra deals with 'unknown' numbers, surely this is algebra: no one knows what number of stones went into that bag.

Algebra is, in fact, often easier than arithmetic. A child is less

67

likely to make a mistake with the bags and stones than with the ordinary operation of the trick, when he thinks of a particular number, say 17, and carries out the successive calculations.

We are now in a position to make up our own tricks. Children enjoy this mildly creative exercise. A little help may be needed to get them started. In music, it is hard if someone says to you, 'Make up a tune.' It is easier if someone says, 'Here is the beginning of a tune. How would you finish it?'

Any beginning whatever can be given. It is always possible to finish it so that it becomes a trick.

In thinking out a new trick it is essential to use the bags and not a particular number. For example, suppose a teacher asks, 'How would you finish this trick? – Think of a number. Add 1. Multiply by 3.'

Suppose a child simply thinks of a particular number, say 2. He adds 1, which gives 3, and multiplies by 3 to get 9. He may then suggest the ending, 'Take away 5. The answer is 4.'

But this does not, in fact, give a trick. It is instructive to see what would happen if different children thought of different numbers.

	Alf	Betty	Charles	Dora	Ernest
Think of a number	0	1	2	3	4
Add 1	1	2	3	4	5
Multiply by 3	3	6	9	12	15
Take away 5	?	1	4	7	10

It is by no means true that the answer is always 4. In fact, the last row strongly suggests that the answer will be 4 only when the number first thought of is 2. There is a pattern in this last row – the numbers rise steadily by 3. So, while we have not invented a trick, we have been led to make an observation, and this might start off a new investigation. Does a pattern of this kind always emerge? The children could make experiments. They could choose some other set of instructions, such as 'Think of a number. Add 4. Multiply by 2. Take away 3', and see what happens in the last row of the table. In the course of this investigation, they will make calculations and improve their arithmetic. It is much more interesting to make calculations in the hope of making a dis-

68

covery than to make calculations because you have been told to.

It often happens in mathematical research that we set out to answer one question, and in the course of the work we observe something that leads us to pose another question. This is one of the ways in which mathematics grows.

To return to our uncompleted trick, we find the simplest way to finish it is to show its progress pictorially (Figure 23).

Think of a number

Add 1

Multiply by 3

Figure 23

A rapid way to finish this trick would be as in Figure 24.

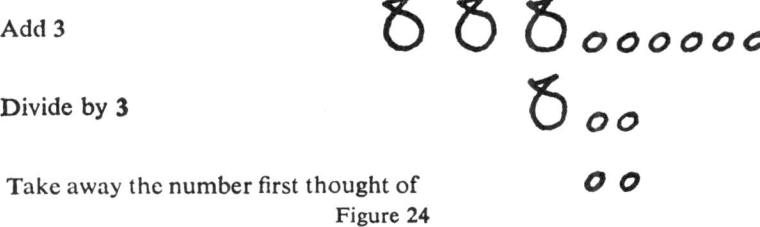

Add 3

Divide by 3

Take away the number first thought of

Figure 24

There are 2 stones left. So the answer is bound to be 2, whatever the number first thought of.

A trick of this kind is complete when we have reached a stage where there is no bag present – for the bag represents an unknown number.

The trick ending given above is a very streamlined one. It requires a certain amount of experience and foresight to find such an ending – more foresight than most children will have on their first contact with this work. Children will often produce a rambling conclusion such as the one below (Figure 25) and this is perfectly satisfactory.

Think of a number

Add 1

Multiply by 3

Figure 25

The children may continue as in Figure 26.

Take away the number first thought of

Add 5

Divide by 2

Take away the number first thought of

The answer is 4

Figure 26

Having arrived at this trick pictorially, we test it arithmetically. Our table now looks like this:

	Alf	Betty	Charles	Dora	Ernest
Think of a number	0	1	2	3	4
Add 1	1	2	3	4	5
Multiply by 3	3	6	9	12	15
Take away original number	3	5	7	9	11
Add 5	8	10	12	14	16
Divide by 2	4	5	6	7	8
Take away original number	4	4	4	4	4

The answer indeed is 4. The children may notice that each row has a pattern of its own. In some rows the numbers rise by steps of 1, in some by steps of 2, in some by steps of 3. The size of the

Tricks, Bags, and Machines

step is connected with the number of bags in the picture corresponding to that row. For example, the fifth picture shows 2 bags and 8 stones; in the fifth row of the table the numbers are 8, 10, 12, 14, 16 rising by steps of 2. The children also may notice that the number in Alf's column is the same as the number of stones in the picture.

The more the children notice these things for themselves and the less they are told, the better the lesson will be. Neither the teacher nor the children should worry about how much is noticed nor whether the children remember all the things they observe. The objective at this stage is a very broad one – to show children that in mathematics you can notice things for yourself, that you can invent things. Actually many of the things observed are relevant to further work in mathematics; the abler students will remember them, and will find this background helpful in later years. But we will only confuse the children if we insist on their remembering a mass of details. It enlivens a lesson if pupils are calling out things they have noticed. But the main objective of a lesson should be simple, and the main objective here is to show children how to explain and invent tricks by drawing bags and stones. This is a simple idea and yet it bridges an enormous gulf – from arithmetic to algebra.

The work with tables incidentally is very good for accuracy in elementary arithmetic. There is so much pattern in the Alf – Betty – Charles – Dora – Ernest tables above, that any mistake stands out clearly. Suppose, for example, that Charles had the mistaken belief that $7+5$ was 13. The fifth row would then read 8, 10, 13, 14, 16. This no longer rises evenly by steps of 2, and indeed 13 stands out as the only odd number in a collection of even numbers. Charles has his attention forcibly drawn to the fact that he has learnt his addition tables wrong. Even with a large class, this method can be used to make children aware of mistakes in their thinking.

TOWARDS ABBREVIATIONS

Once the bag and stone idea has been accepted, it develops naturally towards abbreviations. The following is a trick: 'Think

Vision in Elementary Mathematics

of a number. Add 15. Multiply by 4. Subtract 16. Multiply by 25. Subtract 500. Divide by 100. Take away the number you first thought of. The answer is 6.'

If we illustrated this trick, the fifth picture would have to show 100 bags and 1,100 stones – which most certainly we do not want

to draw. But it is easy to find an abbreviation: $100\,x + 1100$

This conveys quite well the idea of 100 bags and 1,100 stones. We bring in this abbreviation quite casually and naturally. Even in the tricks already considered, which only use small numbers,

we are quite glad to write: $2\,x + 8$

instead of drawing the 2 bags and 8 stones in detail. We work with this type of abbreviation until the children are quite used to it. As we draw, we probably speak. Sometimes we read the picture:

$3\,x + 6$ as '3 bags and 6 stones', sometimes we read it as

'3 times the number thought of and 6'.

A final amendment makes the whole business look much more sophisticated. We erase the top and bottom on a drawing of a

bag. Instead of a bag: x we now see: x This is quicker to

draw than a bag; it is less likely to be mistaken for the figure eight: 8. And it looks like the letter x. So we are now writing $3x+6$ for '3 bags and 6 stones'.

Of course, once x has come into the picture, it is clear to everybody that we are doing algebra.

Our tricks have gone through 4 stages – words, pictures, simplified pictures, shorthand. Our first trick is shown in Figure 27 in 4 forms.

If you look at the shorthand column, you will see that we have made several simple calculations in algebra. In fact, we have covered the 4 operations, addition, subtraction, multiplication, and division. The calculations we have made (without being quite aware of the fact) are: add 3 to x; double $x+3$; take 4

Tricks, Bags, and Machines

from $2x+6$; divide $2x+2$ by 2; subtract 1 from $x+1$. These are not difficult calculations, but they are algebra. The child has made these calculations, without the help of any rules, before he even knew he was doing algebra. His own thinking has led him naturally to these results. It was, in fact, in some such way as this that mankind was originally led to the idea of algebra. The ancient Egyptians pictured an unknown number as a heap of stones, and in their work regularly referred to it as 'heap'. A heap of stones, a bag of stones – there is little difference.

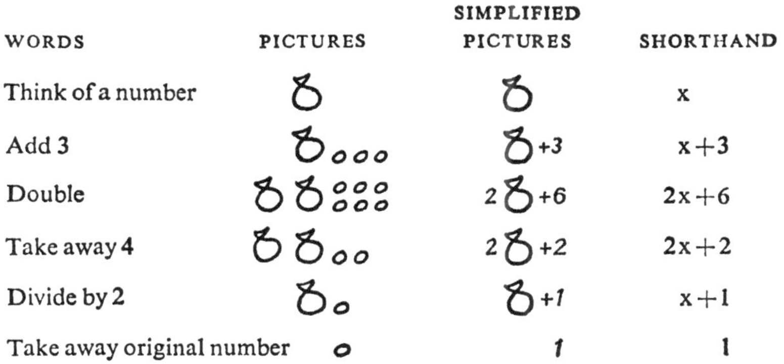

WORDS	PICTURES	SIMPLIFIED PICTURES	SHORTHAND
Think of a number	🜊	🜊	x
Add 3	🜊 ₒₒₒ	🜊+3	$x+3$
Double	🜊🜊 ₒₒₒₒₒₒ	2🜊+6	$2x+6$
Take away 4	🜊🜊 ₒₒ	2🜊+2	$2x+2$
Divide by 2	🜊 ₒ	🜊+1	$x+1$
Take away original number	ₒ	1	1

Figure 27

While children are still interested in this new idea, they may like to practise simple calculations, such as 'Add $2x+3$ to $4x+5$', 'Take $3x+2$ from $7x+6$', 'Multiply $3x+4$ by 5', 'Divide $4x+6$ by 2'. These they will, no doubt, think of in terms of bags – 'If you add 2 bags and 3 stones to 4 bags and 5 stones, you get 6 bags and 8 stones, so the answer is $6x+8$'. Philosophers may worry about this procedure. After all, x is a very abstract idea; should children think of it in such physical terms as 'a bag'? Fortunately, children are generally wiser than philosophers. I should be worried if any of my pupils had ever asked me, 'Is x made of paper, leather, or cloth?' But such a question has never been asked. 'The number 3' is itself an abstract idea, yet children are invariably introduced to it through their experience of actual objects, 3 buttons, 3 trees, 3 cows, 3 aeroplanes. Indeed, there is no other way in which they could learn the idea of 'three'. No different principle is involved in using a bag to suggest an unknown number of stones. And in teaching, we continually remind the children that the bag is there to represent 'the number someone has thought of' or 'any number you may choose', or some other such phrase that the teacher finds suitable.

For example, the bag method of picturing recently led us to the addition result:

$$\begin{array}{r} 2x+3 \\ +4x+5 \\ \hline 6x+8 \end{array}$$

We ask the children, 'How shall we test this?' If they suggest, 'By drawing bags', we ask them what started us off on this business of drawing bags. It was to represent numbers thought of. So we can go back to the Alf – Betty – Charles – Dora – Ernest type of investigation. If Ernest put 4 balls into the bag, for him 2 bags and 3 balls suggests 11 balls, 4 bags and 5 balls suggests 21 balls, and 6 bags and 8 balls suggests 32 balls. So to Ernest the addition above suggests $11+21=32$. If we do the same thing with the others, we arrive at the results:

Alf	Betty	Charles	Dora	Ernest
3	5	7	9	11
5	9	13	17	21
—	—	—	—	—
8	14	20	26	32

All of these are correct additions. They do not prove absolutely that our algebra has been correct, but they make it highly probable.

In words, our conclusion would be stated, 'If you think of any number, and add twice that number and 3 to 4 times that number and 5, you get 6 times that number and 8.' A small, but important point should be noted here. Children tend to read $4x$ as '4 and x', so they associate this symbol with addition rather than multiplication. But it is clear that $4x$, interpreted as 4 bags, implies multiplication rather than addition. If a bag contains a number of stones, 4 bags contain 4 *times* as many, not 4 more stones than a single bag.

One of the wearisome aspects of teaching is that some children have to be reminded of this point not two or three times, but hundreds of times. It is on such points that teaching machines or other mechanical aids may legitimately be used. For there is no new idea involved. The child has understood that 4 bags hold 4 times as many stones as one bag. The trouble is that he does not think of this when he sees $4x$ in a piece of algebra. With some children, a thing has to be repeated many, many times before it lodges in the memory. So some procedure by which a child repeatedly reads a card reminding him of the meaning of $4x$, or by which he is repeatedly tested on such matters, may well be in order. One should take care not to carry this routine drill to the stage at which it produces boredom and disgust.

Tests should vary with the stage the learner has reached. Some children come very quickly to the stage where they can complete a form of the following type*:

Translate into algebra
The number someone has thought of
1 more than that number
1 less than that number
Twice that number
3 times that number
3 added to that number
1 more than twice that number
That number subtracted from 10
5 subtracted from that number

For slower children, the test itself should build up the line of thought required. Teaching machine technique is helpful – in the example below, the learner is supposed to cover the answer column with a piece of paper or a ruler. Immediately after he has written his answer, he pulls the paper, or ruler, down far enough for him to see whether his answer was correct or not.

* The answers are given on page 342.

We have used x as shorthand for the picture ᕷ Give the short-hand for the following pictures:

Picture		Answer
ᕷ ₒ	··········	$x+1$
ᕷ ₒₒₒ	··········	$x+3$
ᕷ ₒₒ	··········	$x+2$
ᕷ ᕷ	··········	$2x$
ᕷ ᕷ ᕷ	··········	$3x$
ᕷ ᕷ ᕷ ₒ	··········	$3x+1$
ᕷ ᕷ ᕷ ᕷ	··········	$4x$
ᕷ ₒₒₒₒ	··········	$x+4$
ᕷ ₒₒₒₒₒ	··········	$x+5$
ᕷ ᕷ ᕷ ᕷ ᕷ	··········	$5x$
ᕷ ᕷ ₒₒₒ	··········	$2x+3$

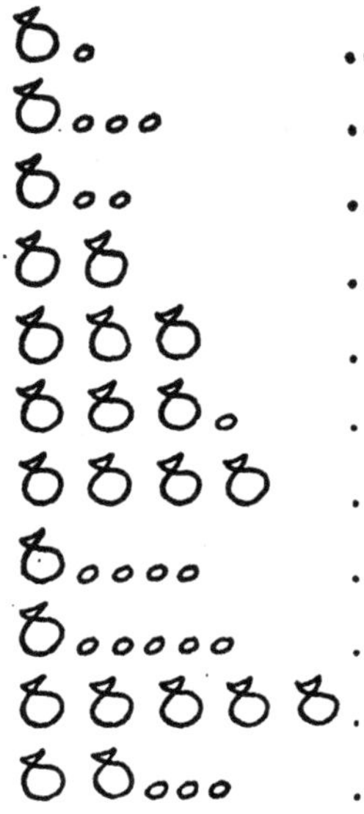

Figure 28

Draw pictures for the shorthand below:

$2x$	················	ᕷ ᕷ
$x+2$	················	ᕷ ₒₒ
$x+3$	················	ᕷ ₒₒₒ
$3x$	················	ᕷ ᕷ ᕷ
$3x+1$	················	ᕷ ᕷ ᕷ ₒ
$2x+1$	················	ᕷ ᕷ ₒ
$x+1$	················	ᕷ ₒ
$4x$	················	ᕷ ᕷ ᕷ ᕷ

Figure 29

Tricks, Bags, and Machines

Draw a picture for $x+2$

If there were 10 stones in the bag, how many stones in your picture altogether?	12
Did you add or multiply?	Add
$x+2$ means that x and 2 are:	Added

Draw a picture for $x+3$.

If there are 10 stones in the bag, how many altogether in your picture?	13
Did you add or multiply?	Add
$x+3$ means that x and 3 are:	Added

Draw a picture of $2x$.

If there are 10 stones in each bag, how many altogether in your picture?	20
Did you add 2 or multiply by 2?	Multiply
Which describes your answer better – 2 added to 10, or 2 times 10?	2 times 10
Complete this sentence. $2x$ means 2 .... x.	2 times x
$3x$ means 3 x	3 times x
$4x$ means 4 x	4 times x
$3+x$ means 3 x	3 added to x
$4+x$ means 4 x	4 added to x
$5x$ means 5 x	5 times x

If x, the number of stones in the bag, happens to be 10, then $3x$ is 30, and:

$4x$ is:	40
$3+x$ is:	13
$4+x$ is:	14
$5x$ is:	50

I am not sure that the above is a very good sample of a teaching machine. An actual teaching machine would probably have far more questions, and indeed would require more space than can be afforded here. But this example does show what a teaching machine tries to do, namely, to put on paper the kind of conversation a good teacher might have with a pupil in difficulty.

The teacher tries to find out what the pupil understands, and gradually to lead him on from there. The pupil is in the dark; he is not sure whether he has grasped the idea the teacher is trying to convey or not; he answers at random and hopes for the best. Every so often he finds he has blundered into the correct answer, perhaps a series of correct answers; the teacher says, 'That's right, that's right, that's right.' From these successful ventures, the student gradually gets his bearings. Of course, there is always the danger of parrot learning; the pupil may have formed no idea at all, but merely have discovered what answers please the teacher. However, we strive to avoid this outcome.

This process, in which the minds of the teacher and the pupil are groping towards each other, can to a certain extent be reduced to routine. A teacher probably has some established idea of the order in which concepts should be communicated; his questioning follows a more or less fixed pattern. A teaching machine simply takes this routine part of the procedure and makes it available to each pupil. The teacher can have a conversation with only one pupil at a time – unless it happens that several pupils meet the same difficulty at the same time. By having his questioning in a permanent printed form, the teacher is freed from a routine part of his duty; all pupils are able to have the benefit of the questioning simultaneously, and each pupil can work at his own speed. The teacher is left free for the more interesting and enterprising parts of his work. He is still needed, of course, to help pupils with their individual and unforeseen difficulties.

A GUESSING GAME

The following is a game that children enjoy. It teaches them both arithmetic and the beginnings of algebra. It can be played equally well in a classroom, or by two people on their own. One person thinks of a rule – for instance, he might choose the very simple rule, 'Whatever number anyone says, I will say 1 more than that'. The others then call out numbers, and the first person answers. This continues until the rule is guessed. A simple rule, like the one mentioned, would very quickly be guessed. We might record the progress of the game as follows:

Number called	7	2	10	100
Number answered	8	3	11	101

By this time someone would have spotted the rule. He might call out, 'The number answered is always 1 more than the number called'; or he might abbreviate this, using x for 'the number called' and y for 'the number answered'. He would then say simply 'y is always $x+1$.' We could write this even more compactly as $y=x+1$.

A few examples are given here to illustrate the variety of possible rules. Answers will be found on page 342.

1. Number called (x)	3	10	$\frac{1}{2}$	98	11	8
Number answered (y)	5	12	$2\frac{1}{2}$	100	13	10
2. Number called (x)	4	2	5	7	10	1
Number answered (y)	6	8	5	3	0	9
3. x	4	10	6	$\frac{1}{2}$	17	$1\frac{1}{2}$
y	8	20	12	1	34	3
4. x	4	7	10	2	3	11
y	40	70	100	20	30	110
5. x	4	7	10	2	3	11
y	41	71	101	21	31	111
6. x	4	10	6	$\frac{1}{2}$	17	$1\frac{1}{2}$
y	9	21	13	2	35	4
7. x	10	14	11	3	4	5
y	5	7	$5\frac{1}{2}$	$1\frac{1}{2}$	2	$2\frac{1}{2}$

In game (2) above, the rule was 'I will subtract the number called from 10'. If someone called 17, the answer might be given, 'I cannot do that. Say a smaller number.' Some children, however, faced with this difficulty, will reply, '7 below zero.'

It is a little hard to guess the rule in the table above for game (6). When the rule is hard to guess, the players will find it wise not to call out numbers at random, but to proceed systematically,

Vision in Elementary Mathematics

calling in turn 0, 1, 2, 3, 4, 5. If they had done this in game (6), the table would have looked like this:

x	0	1	2	3	4	5
y	1	3	5	7	9	11

Here, each time the number called is increased by 1, the number answered is increased by 2. This suggests that doubling comes into it somewhere. If we guess 'You are doubling', that does not work, but in testing this guess we get information that gives us a clue to the correct answer, shown here.

The number called (x)	0	1	2	3	4	5
Double the number called ($2x$)	0	2	4	6	8	10
The number answered (y)	1	3	5	7	9	11

We now notice that the number answered is one more than double the number called; y is one more than $2x$; $y=2x+1$ is the rule.

The art of guessing is to look out for near misses. It is essential to have the courage to make incorrect guesses. Then examine the incorrect guesses to see which seems least incorrect. The person who insists on being right first time has very little chance of solving a problem.

A GUESSING MACHINE

The 'machine' to be described here is not what is technically called a teaching machine. It is a little piece of apparatus that helps children to see the meaning of x and y. Because it is a *thing*, it has an appeal to many children that a verbal problem does not have. You can leave these little objects lying around, and children will ask you, 'What do you do with this?' (See Figure 30).

Its essential idea is exactly the same as that of the guessing game described in the previous section.

Various numbers are written on a piece of cardboard. A pointer can swing round on a nail or drawing pin. One end is marked x, the other y. In the illustration, the number at the y end is always 3 more than the number at the x end. This machine illustrates the rule $y=x+3$.

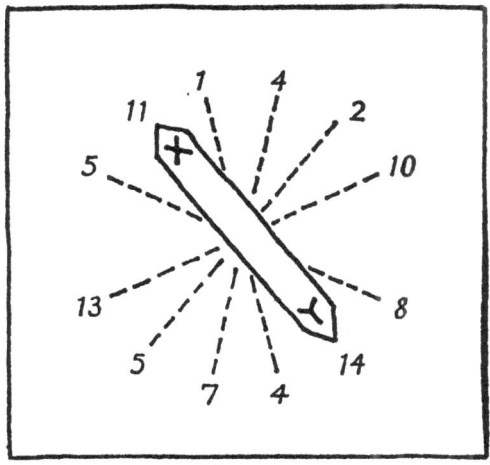

Figure 30

It does not take long to make several little machines, embodying different rules. Handling these machines and guessing the rules, children become familiar with the meaning and use of x and y.

TESTING AND DISCOVERY

In almost any country of the world you can find college students who will cheerfully write down complete nonsense in algebra. I do not mean hasty slips. All of us write nonsense at times, and recognize, when we come back to read it again, that we have done so. But to learn a subject in such a way that you write nonsense and are happy with it and prepared to live with it – this is a strange, though unfortunately, not an unusual, procedure.

The root of the matter is that the student does not feel responsible for what he writes or says. He feels that the teacher is responsible for the truth of the information; the student is only responsible for remembering what the teacher said. In some subjects this may be so. If, for example, a French teacher tells me that the French always use one phrase to convey a particular idea, and never use the phrase that I have put in my homework, I can only bow to superior knowledge. If I produce a dictionary in support of my version, I will be told the dictionary is out of date; people do not speak like that today. Mathematics is, or should be, the most democratic subject in the world; in mathematics there is no evidence available to the teacher that is not available to the student.

The most important thing in the early teaching of mathematics is that the student should form the habit of weighing evidence, of deciding for himself. We are not so much concerned with strict deductive proof as understood by mathematicians. We are more concerned that the student should recognize when some statement is glaringly and outrageously wrong. The student must know when something is sheer nonsense. In algebra, most mistakes are ridiculous mistakes. If the student never writes nonsense, he has a very good chance that his work is correct.

When pupils make an incorrect statement, the teacher should not say, 'That is wrong,' but rather, 'How shall we test that?' If, for example, a child, in studying a 'Think of a number' trick, says 'Twice $x+3$ is $2x+3$', this statement can easily be tested. Here x stands for 'the number thought of'; that number could be, say, 10. Then $x+3$ would mean 13 and $2x+3$ would mean 23. And 23 is not twice 13, so clearly there is something wrong. If such a test seems to indicate that a statement is true, it is well to try one or two other numbers for x. The correct result may have been an accident, due to the particular number thought of. For example, 'twice $x+3$ is $x+6$' would pass the test if we chose the number 0, but it would fail for any other number.

The following example of a child discovering a result in beginning algebra occurred in one of my experimental classes. Some of the abler children were rushing through questions like 'Subtract $2x+3$ from $6x+10$'. In order to keep them busy, while I attended to the slower pupils, I set some questions without any explanation at all. They had to invent a method. One such question was 'Subtract $x-1$ from $2x$'. Their first guess was quite wrong. They wrote $x-1$ as the answer. In the next lesson I asked, 'How shall we test this?' The children suggested three methods of testing the subtraction:

$$\begin{array}{r} 2x \\ x-1 \\ \hline x-1 \quad (?) \end{array}$$

Tricks, Bags, and Machines

1. In arithmetic, they said, we test:

$$\begin{array}{r} 13 \\ 7 \\ \hline 6 \end{array}$$

by adding the two bottom numbers, 7 and 6, and checking that they make the top number, 13. We could do the same with algebra.

2. We could draw pictures. I have 2 bags:  I take

away a bag with 1 stone removed. What remains? (*Note*:

the picture shows a bag with a hole, through which one stone has been removed. We should picture $x-2$ as a bag with 2 holes, through each of which a stone has been removed!)

3. We could try particular numbers for x, 'the number thought of'. We tried 10 for x, so that $2x$ became 20 and $x-1$ became 9. This gave us the unsatisfactory result:

$$\begin{array}{r} 20 \\ 9 \\ \hline 9 \end{array}$$

We now know that, whatever was right, our first guess was wrong. We had recognized a piece of nonsense. So this problem was handed back to the brighter pupils for further study.

NANCY COCHRANE'S METHOD

The correct result was found empirically by a 10-year-old girl called Nancy Cochrane. She used a method which had never occurred to me in connexion with this question. She took several values in turn for x, the number thought of, and performed the subtraction in each case. For example:

> if x is 10
> then $2x$ is 20
> $x-1$ is 9
> Subtraction gives 11

Putting several such calculations together she obtained the table:

x	10	13	15	19
$2x$	20	26	30	38
$x-1$	9	12	14	18
	11	14	16	20

She observed that the answer to the subtraction always seemed to be 1 more than the number first thought of. So she said, 'It looks to me as if the answer should be $x+1$' – which indeed it is.

We then examined her answer by methods (1) and (2). Did it make sense that if I had 2 bags of stones and someone asked for a bag with one stone removed, I should be left with a bag and a stone? It did make sense. Perhaps I gave him a full bag, and he returned one stone to me as change.

We used the pictures to help with method (1) – testing subtraction by addition. Was it correct that $x-1$ and $x+1$ added up to $2x$? Is it true in pictures?

The class looked at this in silence for a moment and then you could almost hear their brains click to the solution; why, of course, you could put the stone into the hole, and get 2 normal bags.

BAGS AND THE BEGINNINGS OF ALGEBRA

When someone is proficient in algebra, he normally solves a problem in the following way. First, he expresses the problem by means of an equation; he solves the equation; then he knows the solution of the problem. Naturally, we want our pupils to reach the stage where they are able to do this. Algebra textbooks usually give some simple exercises designed to lead the pupil in this direction. Unfortunately, children usually do not respond to these exercises in the way they are supposed to. For example, an exercise may read as follows: 'Tommy had 7 marbles. He won some marbles, and then had 9 marbles. Write an equation and solve it to find out how many he won. (Hint: let x stand for the number he won.)'

What children are supposed to do is the following. They should say, 'After winning x marbles, Tommy had $7+x$ marbles. But we are told he had 9 marbles in the end. So $7+x$ and 9 must be the same number. We write the equation $7+x=9$. Therefore, $x=2$. Answer: he won 2 marbles.'

Unfortunately, they do not do anything like that at all. They say, 'Tommy began with 7 marbles and ended with 9 marbles, so he won 2 marbles.' Then you have to point out to them that the book said, 'Write an equation', and we have not written any equation. Rather sheepishly you insist that we must write an equation. This makes the whole business artificial. We are no longer seeking an equation in order to solve a problem. We have solved the problem, and are only writing an equation to please the teacher, or the author of the textbook.

Moreover, the work of writing the equation does not go too smoothly. Tommy had 7 marbles; he won x marbles; how many had he then? You want them to say $7+x$ but they do not. They answer cheerfully, 'Nine' – and after all, the question does say he had 9. The children feel that 9 is a much more definite and satisfactory answer altogether than $7+x$, with its unknown x.

These exercises in fact do not follow the natural flow of a child's thought – or for that matter of anybody else's thought. To this extent, they are intellectually dishonest. No one, in fact, is stupid enough to solve such a problem by algebra. These exercises are not evidence of the power of algebra. So we find ourselves continually forcing the children into an unnatural way of thinking, a way which they quite rightly feel is inappropriate to such problems, and which, in fact, can only be justified by its use in more advanced problems, the existence and nature of which are completely unknown to the children.

The obvious remedy would be to start with a problem that could not be solved by common sense without the help of algebra. The children would then see the power of algebra in action. It is easy to see why writers tend to avoid this remedy. Such a problem would tend to be complicated, and children would lose heart because the exercise appeared so difficult. One has to find a middle region between problems which are so easy that they can be solved without algebra, and problems which appear so hard that the average pupil surrenders at sight of them.

Some of the problems at the beginning of Chapter 3, about the heights of a man and his sons, probably lie in the correct region of difficulty. Solved first by the drawing of pictures, and later by means of equations, they are not too difficult. They may be too easy. Some children will arrive at the answer immediately by pure guesswork. If so, these problems can be made a little more complicated by bringing in larger numbers. A child is less likely to solve $m+s=73$, $m+2s=100$ by sheer guessing. These simultaneous equations represent a problem for which algebra – or the disguised algebra of picture drawing – is a natural method. This is not true of single equations. Many children can solve the problem, 'I think of a number, multiply it by 4, add 3, and get 31' by working backwards, without ever bringing in the equation $4x+3=31$. It is for this reason that this book uses the unorthodox idea of presenting simultaneous equations before single equations.

There are, of course, other methods of approach. We may seek for problems that use only one equation, but are a little more complicated than the one about Tommy's marbles. Consider the following problem. 'Alf has some marbles, Bill has twice as many as Alf, Chris has 3 times as many as Alf. Altogether, they have 60 marbles. How many has each boy?' It is easy to see why books do not begin with this type of question. The sheer mass of words would defeat many pupils. So it is essential to begin by reading through the problem, and drawing it as we go. The first sentence is, 'Alf has some marbles.' It does not say how many, just 'some'; so we visualize Alf as the owner of a bag of marbles. Bill owns twice as many – 2 bags surely is the appropriate picture. In the same way, we picture Chris with 3 bags. The first sentence has now reached diagrammatic form:

Alf   *Bill*   *Chris*

Now we come to the second sentence – together they own 60 marbles. But the diagram shows their total wealth to be 6 bags of marbles. So each bag must hold 10 marbles.

It is not difficult to pass from this argument to the symbolic form, in which Alf has x, Bill has $2x$, and Chris has $3x$, total $6x$, so $6x=60$ and $x=10$.

At a later stage the children may solve such problems by algebra without ever drawing the bags. But when in difficulty, they will always be able to go back to the pictures. Incredible as it may seem to mature mathematicians, even an able pupil may take a long time before he is quite sure that the sum of x and $2x$ and $3x$ is $6x$. He may, for example, get $5x$ by adding the 2 and the 3. The picture will make clear that, even though x does not have any number written in front of it, yet it corresponds to *one* bag, so that the total number of bags is $1+2+3$. Incidentally, the error made by a pupil who writes:

$$
\begin{aligned}
x \\
2x \\
3x \\
\hline
5x \quad (?)
\end{aligned}
$$

is typical of the mistakes made by rote learners. For this pupil is going by what is *written on the paper*; he is not visualizing *what the written symbols signify*.

POWER OF THE VISUAL APPROACH

Illustrating problems by drawing bags is a fairly straightforward procedure. It does not call for any great originality on the part of the person making the drawings. Yet it does enable us to solve puzzles of reasonable complexity. A child who is willing to proceed step by step, drawing pictures as he goes, can solve a puzzle such as the following: 'Dan has some marbles. Joe has one more than twice as many as Dan. Fred has 4 more than 3 times as many as Dan. Altogether they have 23 marbles. How many has Dan?' (One could easily compose much more formidable problems that would still be straightforward to handle.) The solution, by pictures and algebra, would appear as in Figure 31.

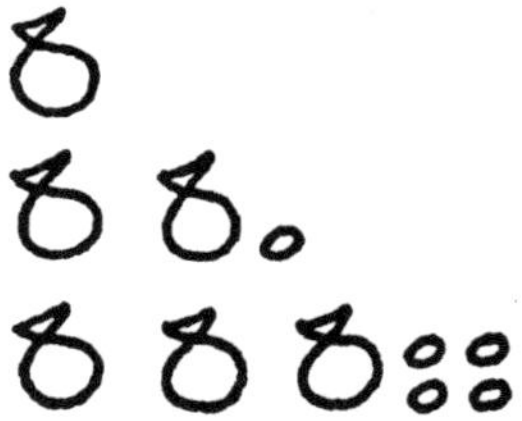

	PICTURES	ALGEBRA
Dan	🗍	x
Joe	🗍 🗍∘	$2x+1$
Fred	🗍 🗍 🗍∘∘	$3x+4$

Altogether	6 bags, 5 marbles	$6x+5$
This is to be	23 marbles	$=23$
	So 6 bags must hold 18 marbles	$\therefore 6x=18$
	So each bag holds 3 marbles	$\therefore x=3$
	Answer: Dan has 3	

Figure 31

CHAPTER FIVE

Words, Signs, and Pictures

ONE of the classic stories in education dates from about 1910. An official visitor to a school in England asked the pupils the following question, 'A shepherd had 80 sheep. 16 died. How many were left?' A quarter of the class added 16 to 80, a quarter subtracted 16 from 80, a quarter multiplied 80 by 16, and a quarter divided 80 by 16. The mechanical part of the calculation was perfect; the children obtained correct answers so far as carrying out the operations was concerned. But it is clear they had not the slightest understanding of what they were doing.

It is a long time since 1910, and one might think that this type of rote teaching has been dead for many years. Yet it is probable that there are many schools in the world today where exactly the same kind of teaching persists. In my own experimental teaching I have often been surprised to meet bright children who had no picture at all of multiplication and division. In one class, all the children knew that $24 \div 4$ was 6. I asked them, 'When would you use this result?' One boy said, 'You might be doing a sum at school' – this was an honest, revealing answer. When else? Deep thought, and then one boy volunteered, 'Your mother might send you to the shop for $24 \div 4$ apples.' I said, 'My mother never did that to me.' The discussion petered out without any fruitful result. I found this surprising. I had imagined that division was always linked with 'sharing' or some similar idea. Apparently, it is still not so.

Efforts are certainly being made to improve the situation. The Scott Foresman series, for example, with the excellent title *Seeing Through Arithmetic*, goes to the lengths of dividing every question into parts. The first part asks, in effect, 'Is this an addition, subtraction, multiplication, or division situation?' Only after this has been answered correctly does the pupil come to the second part of the question, involving the details of the calculation.

In England there has been a widespread movement to make

teaching meaningful. I am indebted to Miss E. E. Biggs for the following essay by a ten-year-old girl:

Story – Arithmetic
by Wendy Smith, aged 10

At the end of every sum we write a story, no matter what kind of sum, add, subtract, multiply, or divide. It has to be a sensible story which happens in everyday life, to describe the number we have used. Just setting numbers down and numbers mean nothing, but writing a story makes them mean something. For example, $46-8+7=45$, that means nothing, but if I write a story and say that 46 people were on a bus, 8 people got off and 7 people got on the bus and then there were 45 people then on that bus, that means something. We have had some very funny examples in our class sometimes. One pupil said two children weighed themselves on a scale, one weighed 3 lb. 8 oz. and the other child weighed 4 lb. 9 oz., altogether they weighed 8 lb. 1 oz., and our teacher wrote on this persons book, what clever babies! So if you write a story like that you cannot understand the meaning of the sum. You know how to do the sum but you do not know why you did it. Two years ago it would just be another sum in the mechanical book 4, but now that sum has a meaning to us. If you know that you have to write a story at the end of the sum, you are thinking of the story as you do the sum, so it makes sense.

Besides writing stories about sums, sometimes we also explain how we have done a sum, and do the sum in as many different ways as we can.

The method of teaching described in this essay has much in common with a device developed independently by my colleague Robert Wirtz in Pennsylvania. For a result such as $7+3=10$, the children are asked to provide a 'Headline' – that is to say, a story illustrating the result. Interesting and varied fantasies result, such as: 7 men took a rocket to the moon. After landing, they found 3 stowaways. So there were 10 men on the moon.

In theory it has been recognized for years, indeed for centuries, that children should understand and not merely parrot arithmetic. In practice, teaching for understanding is still far from universal. It therefore seems worth while to discuss ways of picturing the operations of arithmetic. These pictures will be helpful in learning arithmetic, and they are highly relevant to the beginnings of algebra. Algebra is concerned with the methods of arithmetic. When we teach arithmetic, whether we realize it or not, our real aim is to teach algebra. For it is extremely unlikely that our children will ever meet in later life the exact numbers they had in any problem at school. When we give them any particular exercise, our hope is that they will see that the same method could be applied to many similar problems. So that even when we are dealing with the *particular* numbers of arithmetic, we are hoping to convey *general* ideas, which belong to algebra.

It may be worth while to digress for a moment on this subject of algebra as expressing the methods of arithmetic. In one of my experimental classes, the first crisis arose when we came to the question, 'If there are b boys and g girls in a classroom, how many children are there altogether?' My pupils were most unhappy with the answer, $b+g$. One reason was probably the presence of the plus sign. In all their years of doing arithmetic, the answer had always been *a single number*. They felt that the answer to 'How many children?' should be an honest number, like 23 or 37.

Now, of course, they were quite right in feeling that algebra did not give a definite answer in the way that arithmetic did. The algebra question is essentially concerned with the arithmetical *method*. In effect it says, 'There are some boys in a room and some girls. I do not know how many of each, but that information will reach me tomorrow. I have a calculating machine, which adds, subtracts, multiplies, and divides reliably, but I am stupid, I never know which to do. Can you tell me now what I ought to do tomorrow to find the number of children?' The answer is, 'You should add the number of boys to the number of girls.' This answer can be given in a shortened form. If we adopt b as a convenient abbreviation for 'the number of boys', and g for 'the number of girls', then 'the number of boys added to the number of girls' will appear as $b+g$.

Very likely, in this particular instance, I caused the puzzlement of the class by not appreciating in advance the mental readjustment that this new type of answer called for. Perhaps I could have eased the progress of the class if I had begun by emphasizing that algebra was here a convenient way of describing the methods of arithmetic, and that it did not give an 'answer', a definite number, in the way they had been used to in arithmetic. Once they had grasped this point, we proceeded to make up exercises of our own, eventually reaching such complex questions as 'If there are *c* cats, *d* dogs, *m* men, and *h* hens, how many heads are there altogether? – How many legs? – How many wings – How many ears?' (We decided that hens had no ears – by 'ears' we meant ears hanging down or sticking out from the head, ears that could be seen.) We answered such questions by taking particular examples, working them, and then studying the method used. For example, if there were 5 cats, there would be 20 cat-legs; how did we get 20? – it was 4 times 5. How did the 5 come to be there? – it was the number of cats, *c*. So the number of cat-legs was 4*c*. This was later added to the number of dog-legs, 4*d*, of men-legs, 2*m*, and hen-legs 2*h*, giving the total number of legs as $4c+4d+2m+2h$. The number of wings, of course, is simply 2*h*, twice the number of hens, since each hen has 2 wings, and nobody else has any.

Since algebra gives a very compact way of writing the methods of arithmetic, it can be introduced into arithmetic lessons to make their meaning explicit. For example, suppose a pupil has been able to answer these questions:

How many inches in 2 feet?	24
How many inches in 3 feet?	36
How many inches in 5 feet?	60

The teacher may then lead the student to state the method explicitly. What did we do to 3 to get 36? We multiplied by 12. Did we do the same to 5 to get 60? Yes. Would you do the same with any number? Yes. What do you do then to find how many inches there are in any number of feet I choose to mention? Find 12 times that number. What is a short way to say 'any number of feet I choose to mention'? *x* feet. How many inches, then, are there in *x* feet? 12*x*.

It is essential that the pupil tell the teacher, not the teacher tell the pupil. If the teacher tells the pupil, then it is likely that '*x* feet contain 12*x* inches' will be just another piece of meaningless jargon, added to the confusion already in the pupil's mind. The procedure is only of value if the pupil is already clear as to the meaning of x, and is using algebra as a language to express something in arithmetic which also he understands clearly. In these conditions, and only in these conditions, the algebra will serve to fix and to make explicit the method that is implicit in the particular arithmetical examples.

Another approach to the same question would be to make a little machine, similar to the one shown on page 81, with F and I on the ends of the pointer. Here I is the number of inches in F feet. The pupils would see that the machine had the rule, $I=12F$.

PICTURES OF THE OPERATIONS

We return from this digression on the way in which algebra filters into arithmetic, to consider pictures for the basic operations of arithmetic.

Addition. Addition is easy to picture. We have already had one illustration of it, when the boy climbed on to his father's head (see page 41). Addition is associated in our minds with 'putting together'. We may represent the addition of three and two by a picture such as this:

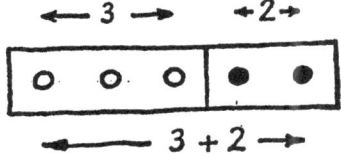

When we are dealing with larger numbers, it may be inconvenient to draw all the objects. Something is left to the imagination as in the illustration of $58+23$:

And this leads on quite naturally to the representation of an 'unknown number':

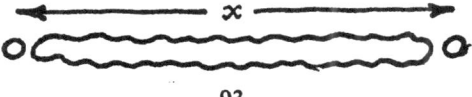

Here we have some objects, but we cannot see just how many there are: a cloud has come in front of them (as in Chapter 1).

We can show $x+3$ like this:

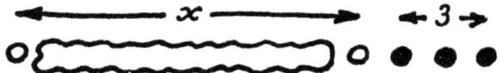

and $x+y$ like this:

In these pictures of addition, we have drawn or indicated a number of objects. This type of picture is, therefore, most natural when we are dealing with whole numbers. We might use it to show a simple fraction, for example:

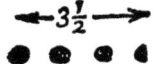

However, the picture of $x+y$ above suggests very strongly that x and y represent whole numbers. This is not a serious inconvenience. If a child can learn to use and think correctly about x and y, with the agreement that 'the numbers thought of' must be whole numbers, little difficulty will be found at a later stage, when x may stand for $3\frac{1}{2}$ or $2\frac{3}{4}$.

Subtraction. We have already had one picture of subtraction – on page 54 where the man is holding his son upside down. The most obvious picture comes from the kindergarten explanation of subtraction, 'taking away'. Our picture of $5-3$ would thus be 5 objects from which 3 had been removed.

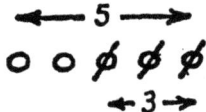

The lines are supposed to show that 3 objects have been crossed out.

Thus $x-y$ would have a picture of x objects, of which y have been crossed out.

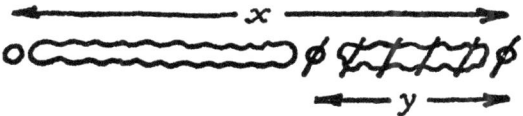

Books on the teaching of arithmetic usually emphasize, and quite rightly, that a subtraction question can be posed in many different forms. For example, $5-3$ may appear under any of the following forms:

> If you take 3 from 5, what is left?
> How much bigger than 3 is 5?
> What must be added to 3 to give 5?

If we suppose a man to be 5 feet high, and his son to be 3 feet high, we can translate these into the following graphic forms:

If you cut off the man a piece as long as his son, how much remains? (Diagrams A and B in Figure 32).

How much taller is the man than his son? (Diagram C in Figure 32.)

On how high a stool must the son stand to appear as tall as his father? (Diagram D in Figure 32.)

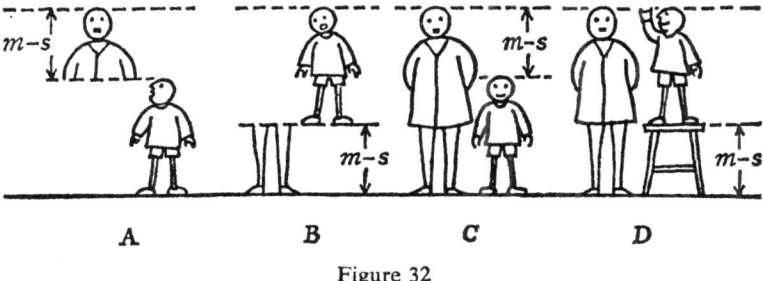

Figure 32

These ways of looking at subtraction can prove useful at a later stage in the pupil's life, when he reaches negative numbers. If he is asked to subtract -5 from 7, it may help him to re-state the question in either of the following forms:

How much hotter is a temperature of 7° than one of −5°? If the temperature is −5°, by how much must it rise to become 7°? Either question may be illustrated by Figure 33.

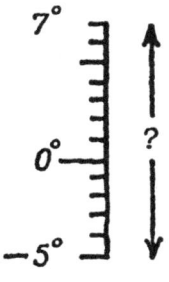

Figure 33

This approach works equally well when 2 negative numbers are involved. To subtract −8 from −2 we may ask 'How much hotter is −2° than −8°?' or 'If the temperature is −8°, by how much must it rise to become −2°?' The picture is shown in Figure 34.

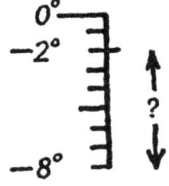

Figure 34

Children who find difficulty with arithmetic may find that this does not solve their problems, because they cannot remember which way round the question is to be put. A large number of errors in mathematical work are due to this difficulty, that the pupil cannot remember what the question means. In elementary arithmetic, the difficulty does not arise, because the smaller number is always taken from the larger. If a subtraction is concerned with 8 and 5, the answer (at this stage) must be 3. But later on, when negative numbers are envisaged, the pupil has to look carefully and see whether the question is '5 from 8' or

'8 from 5'. Still, the difficulty is not acute. Children can still see that £8 from £5 is something that is only made possible by the credit system; a debt of £3 is an intelligible enough answer. It is when we come to take 3 from -4, or -4 from -6, or -7 from 1 that the real trouble begins. At this stage it becomes necessary to link the 'taking away' picture of subtraction with some other image. In the early stages of arithmetic a child should come to realize that '7 from 12' has the same answer as '7 and what are 12?' For this allows him to apply his knowledge of the addition table to carry out subtractions. When we come to negative numbers the 'and what?' approach will still work. The question 'A debt of £4 and what makes a debt of £6?' still makes sense, and leads to the answer, 'A debt of £2'. And this leads to the answer approved by mathematicians: -4 from -6 is -2. The question of negative numbers will be taken up later in more detail.

Multiplication. In Chapter 1 we saw that 3×4 could conveniently be represented by a rectangle (Figure 35).

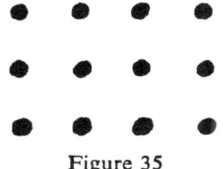

Figure 35

This simple device for representing multiplication proves extremely valuable in many parts of arithmetic and algebra. We may use it to represent $3x$ as in Figure 36.

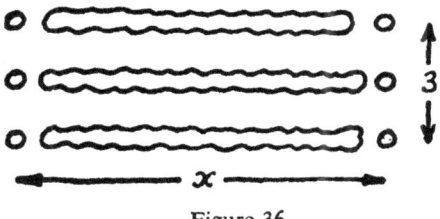

Figure 36

xy may appear as in Figure 37.

97

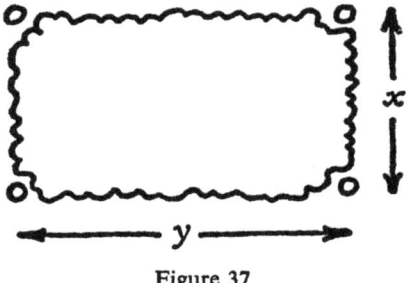

Figure 37

This is intended to represent x rows with y objects in each, the actual numbers being hidden by clouds.

Many useful results can be illustrated by little jigsaw puzzles, in which such rectangles are fitted together.

Another device that is sometimes useful for representing multiplication is the tree.

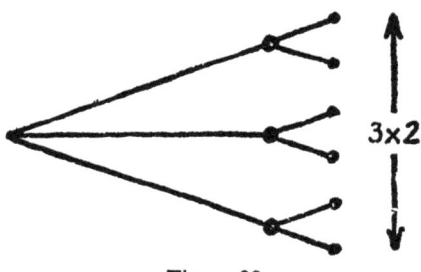

Figure 38

In Figure 38, we have 3 main branches, from each of which 2 twigs grow. Thus we see 3 twos at the ends of the twigs.

'Standing at the crossroads' is a metaphor for making decisions. The tree is often used in problems concerned with successive decisions. If there are 3 actors A, B, C and two actresses, X, Y, available for a little sketch involving 1 man and 1 woman, how many possible casts do I have to consider?

3 roads lie before me when I consider whether to choose A, B, or C for the man; whichever road I take, it forks into 2 when I decide whether to choose X or Y for the woman.

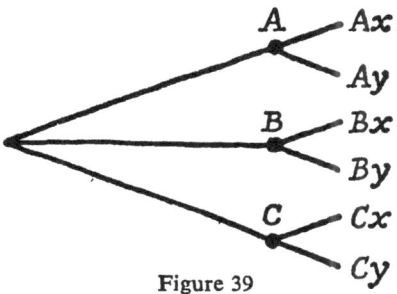

Figure 39

Questions of this kind arise in connexion with football pools and other competitions, when one tries to decide how much it would cost to cover all possible answers.

In the Japanese Hiragana alphabet, each character represents a sound such as *ka, ru, si,* or *o.* There are 10 possible beginnings – the consonants k, s, t, n, h, m, y, r, w, and 'blank'; by 'blank' I mean that the sound has no initial consonant (see first row below). The ending is always one of the five sounds, a, e, i, o, u. How many characters are there in this alphabet? We could show this as a tree with 10 branches and 5 twigs on each. We choose a branch when we select the beginning of a word, a twig when we decide on the ending. Thus there are 50 characters in all.

This problem could equally well be solved by the rectangle illustration. We can arrange the possible Japanese sounds systematically in a rectangle, like this:

| | Endings | | | | |
Beginnings	A	E	I	O	U
'Blank'	a	e	i	o	u
K	ka	ke	ki	ko	ku
S	sa	se	si	so	su
T	ta	te	ti	to	tu
N	na	ne	ni	no	nu
H	ha	he	hi	ho	hu
M	ma	me	mi	mo	mu
Y	ya	ye	yi	yo	yu
R	ra	re	ri	ro	ru
W	wa	we	wi	wo	wu

Without actually going to the trouble of writing all possible combinations, one could see from the rectangle arrangement, 10 rows of 5, that 50 characters would arise in this way.

Division. It is usually emphasized, when arithmetic teachers are being trained, that division can arise in two different forms. These correspond to the two questions, 'If a 60-inch rod is cut into 3 parts, how long will each be?' and 'How many 3-inch rods can you cut from a 60-inch rod?'

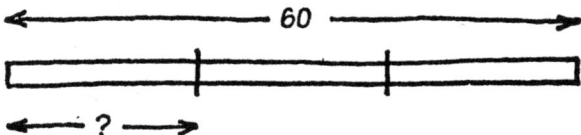

The above picture illustrates the first question. It does indeed appear different from the situation envisaged in the second question, in which we are trying to see how many 3-inch pieces make up a 60-inch piece.

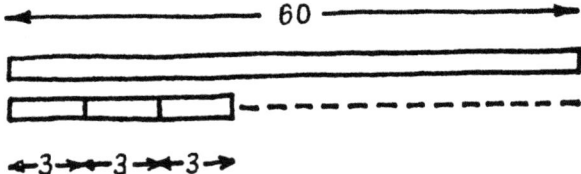

The reason why 20 figures in both answers is that order is unimportant in multiplication. In the first picture above we see 60 as composed of 3 twenties; in the second, we see it as 20 threes. The first question asks, '3 times what is 60?' The second asks, 'What times 3 is 60?' It is not surprising that the answer is the same.

If children are familiar with these two forms of the division question, it may help them when they come to that thorny question, division of fractions. If they meet the expression $60 \div \frac{1}{2}$, they may well be able to answer the question in the form, 'What times $\frac{1}{2}$ is 60?' This is by no means difficult to think out. But they may find it easier to answer the question in the alternative form, '$\frac{1}{2}$ of what is 60?'

The value of the latter form appears more strongly when the fraction is more complicated. A child should be able to think out the question, 'How many $\frac{3}{4}$ are there in 60?' For $\frac{3}{4}$ taken 4 times will make 3, and there are 20 threes in 60, so there must be 80 $\frac{3}{4}$ in 60. But it is easier to answer the question, '$\frac{3}{4}$ of what is 60?'

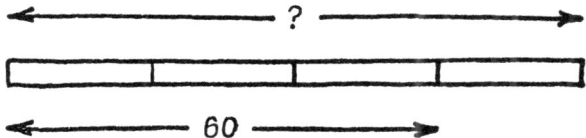

This picture shows how we visualize the problem. The unknown length is broken into 4 quarters; 3 of these are to equal 60. So each piece must be 20, and the unknown length must be 80. No rule has been used here, but those with memories of 'invert and multiply' may be interested to observe that we have carried out exactly the steps that rule commands. For, to find the size of each piece we first divided 60 by 3; then, to find the whole length we multiplied by 4. This is exactly what we would do if we had used the rule and calculated $60 \times \frac{4}{3}$.

BRACKETS

A language problem arises in a restaurant when a waiter shouts to the kitchen, '2 egg and bacon'. Does he mean that one customer wants 2 eggs with his bacon, or that there are 2 customers, each requiring egg and bacon? Sometimes it is agreed between the waiter and the cook that the latter situation be conveyed by 'Egg and bacon, twice'.

This problem appears in arithmetic when we meet expressions of the form $2 \times 3 + 4$. An agreed convention regulates the use of such expressions, but someone ignorant of this convention might be in doubt whether 10 or 14 was intended. The dilemma is that of the restaurant, with 3 instead of egg and 4 instead of bacon. The mathematical convention is that $2 \times 3 + 4$ means 2 threes and a single 4. If we want to convey that 3 is added to 4, and the whole result is to be taken twice, we write $2 \times (3 + 4)$. The pictures in

Figure 40 show what we expect to be served with, when we issue these two orders.

$$2 \times 3 + 4 \qquad\qquad 2 \times (3 + 4)$$

Figure 40

Such pictures save a lot of verbiage. A verbal explanation tends to be in terms of how you calculate $2 \times 3 + 4$; we find ourselves saying, 'You must do the multiplication before you do the addition', and the pupil comes to think more and more of arithmetic as something that we *do* on paper with a pencil. But the proper question to ask in arithmetic is not 'What shall I do to this?' but rather 'What is the situation?' Our first aim should be to see something clearly; only after that does the question arise of what action we are to take. Our pictures try to show what a mathematician means when he writes $2 \times 3 + 4$ and $2 \times (3 + 4)$.

These pictures can be used equally well when unknown numbers are involved. We can show the distinction between $3x + 2$ and $3(x + 2)$. (See Figure 41.)

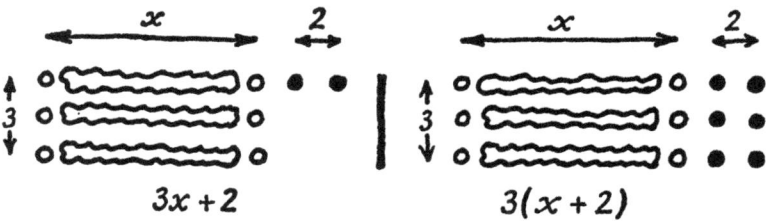

$$3x + 2 \qquad\qquad 3(x + 2)$$

Figure 41

This distinction could, of course, be equally well pictured in terms of bags and stones, $3x + 2$ denoting 3 bags and 2 stones, while $3(x + 2)$ meant a bag and 2 stones, all of this 3 times.

The words 'all of' are a useful way of reading brackets. Thus $2 \times 3 + 4$ could be read '2 threes and four' but $2 \times (3 + 4)$ would be read 'twice all-of-3-and-4'.

The expression $(6+2) \times (3+4)$ could thus be read 'all-of-6-and-2 *times* all-of-3-and-4', that is to say, 8×7 (see Figure 42).

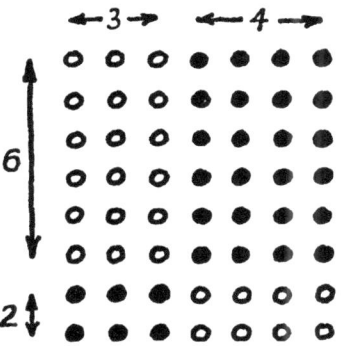

Figure 42

It might not be obvious to someone who began with this picture and wanted to write an expression for it, that the two brackets were necessary. But in the rows we see $3+4$ objects, *all of them* repeated in each row, and in the column $6+2$ objects, *all of them* occurring in each column.

If either bracket were omitted the meaning would be changed. $(6+2) \times 3+4$ would mean: 'All-of-6-and-2, 3 times; and 4' (see Figure 43).

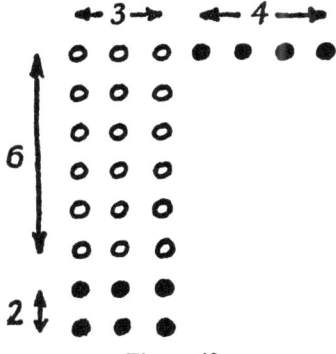

Figure 43

103

$6+2\times(3+4)$ would mean '6 and twice all-of-3-and-4' (see Figure 44).

Figure 44

Curiously enough, the distinction to be made here appears much more easily in algebra than in arithmetic. In Chapter 3, it was quite natural to use $4m+3s$ in describing the height of 4 men and 3 sons. The language of algebra here runs very close to the language of everyday life. Both in algebra and in ordinary language, we have to make a little effort to express the other idea – 4 teams, each consisting of a man and 3 sons, which resembles $4(m+3s)$ in algebra. In English we have to think of a word such as 'team'; in algebra, we use brackets. The team ties all its members together into a new, collective unit; the bracket ties together all the numbers written inside it.

In arithmetic, the natural meaning of $4m+3s$ tends to be obscured. If each man is 5 feet high and each son is 2 feet, $4m+3s$ represents the number $4\times5+3\times2$. This still means '4 fives and 3 twos' but it does not stand out so clearly. We could sympathize with a pupil who looked at the middle of this expression, saw $5+3$, thought, '$5+3$ is 8', and replaced the expression by $4\times8\times2$, – which, of course, is entirely incorrect. The mistake is not one of mathematics but of language. The pupil has not appreciated the idea the person was trying to convey when he wrote $4\times5+3\times2$. If this had been spoken in the form '4 fives and 3 twos', the mistake would probably never have arisen.

We have so far discussed the expression $(6+2)\times(3+4)$ in terms of the situation it describes, the picture it implies. This has been emphasized, because it is an aspect so often neglected. However, we should not go from one extreme to the other, and overlook the fact that this expression can be interpreted as an instruction to calculate. It does tell us what to do with our paper and pencil. In a concise form it specifies the following procedure. (i) Add 6 and 2. (ii) Add 3 and 4. (iii) *Then* multiply the sums found in steps (i) and (ii).

With children one can use the image that the brackets, ( ), are like the blinkers on a horse. We can see nothing outside these until we have finished calculating what is inside them. Only after that may we look at the rest of the world (see Figure 45).

Figure 45

Once the meaning of brackets has been understood, a pupil is already in a position to obtain simple results in algebra. For example, suppose we are concerned with $(x+2)$ $(y+3)$. The pupil will probably need to be reminded that the multiplication sign is omitted in algebra; $(x+2)$ $(y+3)$ means 'all of $x+2$' *times* 'all of $y+3$'. We may write it as $(x+2).(y+3)$ or even as $(x+2)\times(y+3)$, so long as we are careful not to confuse x and $\times$. We picture this product as a rectangle, $x+2$ rows with $y+3$ objects in each (see Figure 46).

This picture quite naturally falls into 4 parts. In the north-west corner, we have a rectangle, x rows of y objects. This rectangle contains xy objects. To the north east we see 3 columns each containing x objects; so $3x$ objects are here. In the south west, we see two rows of y objects; these contain $2y$ objects. To the south east, we see a little rectangle containing 6 objects.

Thus our original rectangle, drawn to illustrate $(x+2)$ $(y+3)$ objects, breaks up into 4 parts containing the following numbers of objects:

xy	$3x$
$2y$	6

105

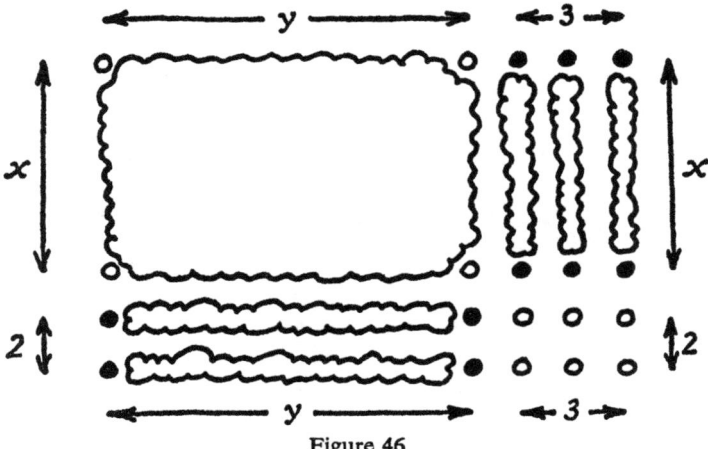

Figure 46

It follows that.

$$(x+2)(y+3)=xy+3x+2y+6$$

Pupils can check that this is so by choosing particular numbers for x and y.

At present, there seems to be no particular purpose in noting this result. Actually, in algebra we often do require results of this type. For the moment, we are concerned only to point out that it is easy to obtain them whenever they are wanted. Children may enjoy working out a few simple multiplications by drawing pictures – this can indeed be their standard procedure in early work on algebra. One can also make a puzzle, in which we start with the pieces shown in Figure 47 and try to fit them together to make

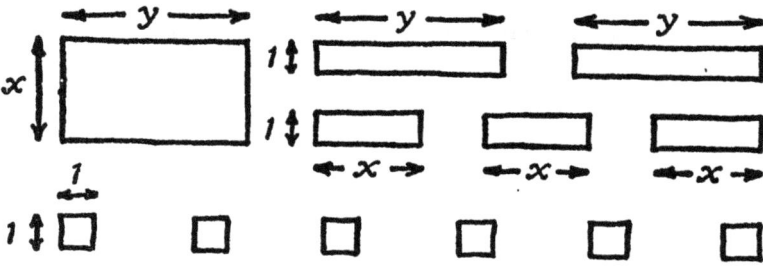

Figure 47

106

Words, Signs, and Pictures

a rectangle. This puzzle is equivalent to asking the pupil to factor* the expression $xy+2y+3x+6$. When he has arranged these pieces in a rectangle, he will see that the sides are $x+2$ and $y+3$.

In arithmetic, it is usually possible to avoid the use of brackets. The expression $(6+2)\times(3+4)$ could arise in a problem of the following kind: 'There were 6 boys and 2 girls at a party. Each of them received 3 large toys and 4 small toys. How many toys altogether did the host give to the children?' In arithmetic we should probably argue: there were 8 children and each received 7 toys. The total number of toys was, therefore, 8×7, or 56. In arithmetic, we do not drag around expressions such as $6+2$ or $3+4$; we work them out and replace them by the numbers 8 and 7. But suppose we want to describe the *method* without mentioning particular numbers. We suppose there are b boys and g girls. The number of children is thus $b+g$, and we cannot 'work this out' or express it in any shorter way (compare page 91). In the same way, if each child receives l large toys and s small ones, this makes $l+s$ toys for each child. To find the total number of toys we must multiply the number of children, $b+g$, by the number of toys each receives, $l+s$. This instruction we convey by means of the expression $(b+g)(l+s)$. (Compare the 'instruction to calculate' on page 104.)

An arithmetic textbook, written purely to teach commercial arithmetic, may not mention brackets at all, on the grounds that brackets are unnecessary for arithmetic. There are other points that can be left out of arithmetic, but which become essential when algebra is started – for example, the convention that $2\times3+4$ means 10 and not 14; expressions involving repeated multiplication, such as $2\times3\times4$; the use of indices,† as in 10^7 or 5^3. All of these are points of language rather than of mathematics. They do not call for reasoning, but for knowing, understanding, and

---

* There are variations between English and American usage. Where there is a choice, the shorter form is generally better. It is ugly and ridiculous to use 'transportation' for 'transport', or 'burglarized' for 'burgled'. The American form, 'factoring', however, is better than 'factorization', used by some English writers — Chrystal, for example.

† Called *exponents* in America.

remembering the meaning of certain expressions. If all these linguistic points are left out of arithmetic, they accumulate at the beginning of algebra, and distract the learner from his main problem, which is to see why letters x and y have made an appearance in mathematics. He will find it much easier to understand what x^3 means if he is already familiar with 2^3; he will find $2(x+4)$ easier if he has met $2 \times (3+4)$. A pupil cannot go far in algebra without meeting the expression $2xy$; this expression is sure to puzzle him if he does not understand the meaning of $2 \times 3 \times 4$. It is, therefore, to be hoped that teachers of arithmetic will mention such things as 2^3 and $2 \times (3+4)$ and $2 \times 3 \times 4$, not for any immediate object in commercial arithmetic, but as a help to those pupils who may later need algebra for higher mathematics, science, engineering, workshop technology, for general reading, or for recreation.

REPEATED MULTIPLICATION

Repeated multiplication is one of the easier things to explain. It is only mentioned here because in experimental teaching I have met children who had never had the idea mentioned to them, and in consequence were puzzled. By $2 \times 3 \times 4$ we understand that 2 is multiplied by 3 (giving 6) and the result is multiplied by 4 (giving 24). Here we have worked from left to right, as in reading. Children may ask; would we get the same answer if we worked from right to left? With $2 \times 3 \times 4$ this would mean that we began with 4, multiplied it by 3, and the result by 2. As 3×4 is 12, and 2×12 is 24, we find the answer is the same. Indeed, we can combine the numbers in any order without affecting the final answer. If we take the ends first, 2×4 is 8 and 3×8 is 24.

It is by no means obvious to children that the order of multiplication is immaterial; indeed, many adults would find it hard if pressed to say *why* we always get the same answer. The fact that the multiplications may be done in any order is important. It is good for children to experiment with a variety of expressions such as $2 \times 5 \times 7$ and $11 \times 13 \times 9$ until they have become satisfied that the final answer is independent of order. Here, as always, we do not *tell* the children the answer. We ask, 'What would the effect of changing the order be?' Different children will have different opinions. We let them experiment and argue until they have reached unanimity.

When everyone has come to feel that order is unimportant, one can produce the picture in Figure 48 that helps to show why this is so.

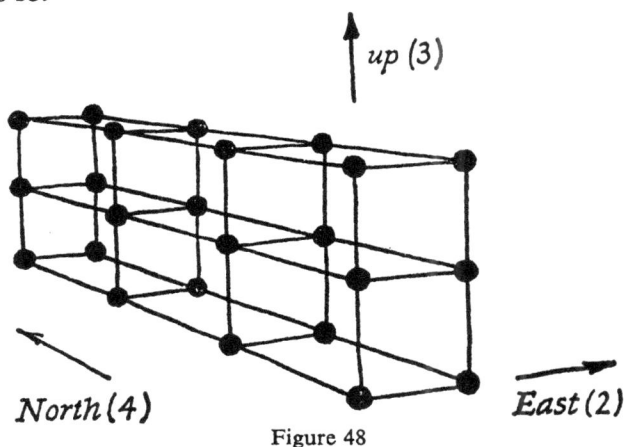

Figure 48

Figure 48 is intended to represent a number of balls mounted on wires. It represents the product $2 \times 3 \times 4$. At the front of the block we see 6 balls, arranged in the pattern we associate with 2×3. As we go north, we meet this set of 6 objects 4 times. Thus, the total number of objects is 6×4, which is what we found when we worked out $2 \times 3 \times 4$, by proceeding from left to right. The eastern face of this block contains 12 balls, arranged in the rectangle we associate with 3×4. So does the western face. Thus, the total number of balls can be seen as 2×12. This product appeared in our earlier work, when we found $2 \times 3 \times 4$ by going from right to left. Finally, the roof of our block contains 8 balls, and we see 8 balls also on the middle floor and on the ground floor. Thus, the total number of balls may be seen as 3×8. But the number of balls in this picture must be the same, by whatever means we calculate it. That is why we found 24 by each method.

An actual model would demonstrate this point even more clearly than a picture.

One could also illustrate $2 \times 3 \times 4$ as being the number of bricks in the block shown in Figure 49.

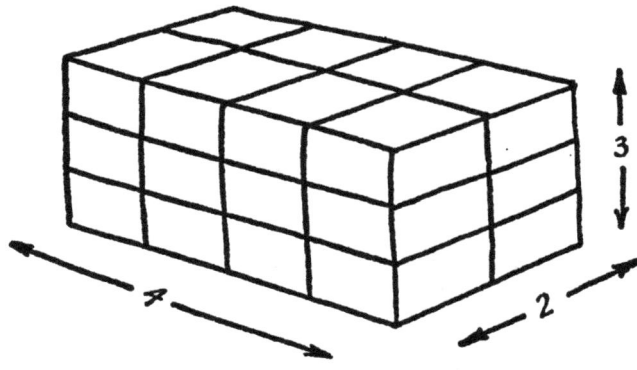

Figure 49

This illustration is all right *if actual bricks are used*. The picture has certain dangers, due to the fact that a child's vision is not always the same as an adult's. I once asked a class of able ten-year-olds how many bricks this picture represented. They answered, 'Twenty-six.' I could not understand how they arrived at this answer. One girl explained it; she had counted the visible squares – 8 blocks on the roof, 6 at the front, 12 at the side, making 26 in all. Her method of counting is shown in Figure 50.

Figure 50

Words, Signs, and Pictures

It will be seen that one brick has been counted three times; the numbers 7, 9, and 18 are all on the same brick. Several bricks have been counted twice, for example, the same brick is marked 1 and 15. On the other hand, several bricks have never been counted at all – the bricks that are out of sight. This misunderstanding was not cleared until an actual set of bricks had been found, a block built, and the bricks in it counted. The difficulty these children found was, no doubt, connected with the fact that they had learnt arithmetic entirely from books, and had insufficient experience of handling materials. There is a general moral in this story – a teacher continually has to remember that something which is clear and obvious to him may have no meaning, or a different meaning, to the children he is teaching. It is necessary all the time to check that communication has been established. And even then we cannot be sure; our imagination may not be able to conceive some of the unexpected ways in which children's thinking differs from ours.

INDICES

The notation of indices takes a long time to establish, and that is why we should begin to teach it early. It is not necessary to lay any great weight on it. A thousand is $10 \times 10 \times 10$; it may be mentioned, rather casually, that some people prefer to write this as 10^3. Such odd references, made from time to time, gradually sink into a child's mind. It is in a similar manner that a child comes to learn a new word in everyday life, by repeatedly hearing it said, but not particularly emphasized.

Emphasis tends to create anxiety and to hinder learning.

Some children confuse 10^3 and 10×3, and it may take many months to cure this confusion. Children hardly ever confuse 10^3 with $10 + 3$. To the question, 'What is 10^3?' one never seems to get the answer 'Thirteen'. But the answer 'Thirty' is common and persistent. Presumably the confusion arises because 3 *tens* are used in writing $10 \times 10 \times 10$; and do not 3 tens make 30?

It might well be wise to approach the subject from another direction. Prime factors, for example, lead quite naturally to indices. If we start with a number such as 72 and ask for its factors, we may get different answers, 8×9 or 6×12 or 2×36 or 3×24 or 4×18. However, these can be broken up further; since $8 = 2 \times 2 \times 2$ and $9 = 3 \times 3$, we can write $72 = 2 \times 2 \times 2 \times 3 \times 3$. If we use the form 6×12 and break 6 up into 3×2 and 12 into $3 \times 2 \times 2$ we find $72 = 3 \times 2 \times 3 \times 2 \times 2$. Here we have a reminder of an earlier result: the order of factors in multiplication does not matter. In each case, the factor 2 occurs 3 times and the factor 3 occurs twice.

Children generally welcome any abbreviation that reduces the labour of writing, and they will accept gladly the suggestion that 2^3 be written as shorthand for $2 \times 2 \times 2$, and 3^2 for 3×3. Thus $72 = 2^3 \times 3^2$ expresses 72 as a product of prime numbers.

It will be found that 2×36 and 3×24 and 4×18 also lead to the same result.

Children do not seem to find the same difficulty translating *into* indices that they find in translating *from* indices. Most children will quite correctly work out $32 = 2 \times 2 \times 2 \times 2 \times 2 = 2^5$, but in the opposite direction many fall into the error of thinking that 2^5 means 10. It would, therefore, seem to be sound to give them plenty of practice with what they do correctly, so that their brains repeatedly register the correct result – 32 is 2^5; 32 is 2^5; 32 is 2^5 – and are not confused by memories of the incorrect answer.

When the translation into indices is well-established, we may begin, cautiously, to ask questions in the opposite direction. This is best done, at first, immediately after a correct piece of translating into indices. If a child has just worked $8 = 2 \times 2 \times 2 = 2^3$, we may ask, 'What number then does 2^3 represent?'

In teaching indices, it is well to start with the larger numbers. Children see some point in writing 3^5 as shorthand for $3 \times 3 \times 3 \times 3 \times 3$. There is not the same obvious purpose in writing 3^2 for 3×3. One can, however, work gradually down:

$$3 \times 3 \times 3 \times 3 \times 3; \quad 5 \text{ factors}; \quad \text{we write } 3^5$$
$$3 \times 3 \times 3 \times 3 \quad ; \quad 4 \text{ factors}; \quad \text{we write } 3^4$$
$$3 \times 3 \times 3 \quad ; \quad 3 \text{ factors}; \quad \text{we write } 3^3$$
$$3 \times 3 \quad ; \quad 2 \text{ factors}; \quad \text{we write } 3^2$$

The exercises on expressing numbers in terms of their prime factors help children to avoid a difficulty of language that might otherwise confront them. If we write $250 = 2 \times 125 = 2 \times 5 \times 5 \times 5 = 2 \times 5^3$, it becomes fairly clear that the index 3 refers only to the factor 5. It has nothing to do with the factor 2. Each factor has its own index, to say how many times it occurs. Thus $7 \times 7 \times 11 \times 11 \times 11$ is $7^2 \times 11^3$. It would be very difficult and inconvenient to devise any other form of notation.

But this is not clear to a child who meets 2×5^3 without the background of such work. A child who is hurriedly introduced to indices will easily fall into the following mistake of interpretation. Asked to calculate 2×5^3, he will *first* find $2 \times 5 = 10$; *then* $10^3 = 1,000$. It will seem to him very arbitrary if we just say, 'You have done the operations in the wrong order. You should first raise 5 to the power 3 and then multiply by 2.' Why this particular convention? On the other hand, the child who has worked with prime factors will more easily see that $1,000 = 10 \times 10 \times 10 = 2 \times 5 \times 2 \times 5 \times 2 \times 5 = 2^3 \times 5^3$, which is not the same as 2×5^3.

In fact, the child who gets the answer 1,000 is working out what mathematicians have agreed to call $(2 \times 5)^3$. Now $(2 \times 5)^3$ and 2×5^3 do not look much different, and it is not surprising that they are sometimes confused. But their meanings are very different; 1,000 is not 250; the cube of twice 5 is not twice the cube of 5.

We can, of course, meet this situation with some sort of rule about the order of operations. But it is better if we can lead the pupil, through the experience of prime factoring, to the most natural system of shorthand. If someone is sufficiently industrious to express 162,000,000 in prime factors, he will find that the factor 2 occurs 7 times, the factor 3 occurs 4 times, and the factor 5 occurs 6 times. So we denote it by $2^7 \times 3^4 \times 5^6$. Each index refers to one factor alone; 7 to the factor 2; 4 to the factor 3; and 6 to the factor 5. This is surely the natural and obvious convention. With this convention, there can be no doubt about the meaning of 2×5^3. It tells us to repeat the factor 5 three times. It does not tell us to repeat 2 at all. So it must represent $2 \times 5 \times 5 \times 5$.

This point is stressed because it causes much misunderstanding in algebra, where we frequently meet expressions such as $2x^3$. If the pupil does not see, with crystal clearness, that this means 2 times x times x times x and *not* $2x$ times $2x$ times $2x$, the whole of his early work on algebra will be a fog of misconceptions.

SQUARES AND CUBES

The product 3×3 has this picture: • • • / • • • / • • •   The shape of a square seen here, and 3×3 is often called 'the square of 3'. The shorthand form, 3^2, is usually read '3 squared'.

If we were to take 3 squares, such as that shown above, and place them one above another, we should obtain a cube. This cube would contain $27 = 3 \times 3 \times 3 = 3^3$ dots. We often speak of 27 as being the cube of 3.

The same type of picture is possible when we are dealing with an unknown number x. We can picture x^2 by means of x rows containing x dots in each. We picture x^3 by means of a cube, having x layers, and each layer containing x rows of x dots (see Figure 51).

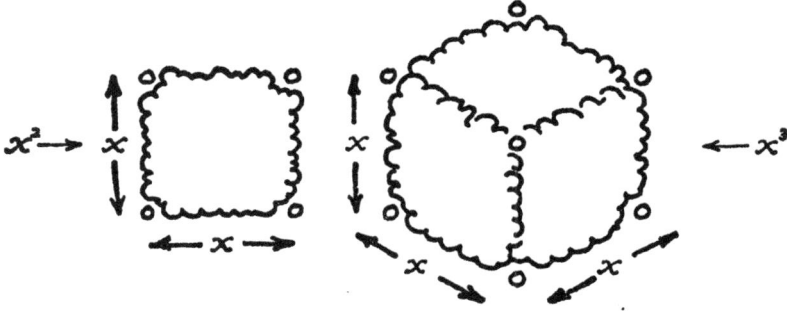

Figure 51

As always with x, there is plenty of cloud about, which prevents us from seeing just how many dots there are in each row. The clouded cube is a peculiar-looking object, and it may be convenient at times to draw it simply as a box (see Figure 52).

It must be emphasized that this box is packed with objects, perhaps tennis balls. The box is full of these. The picture is not

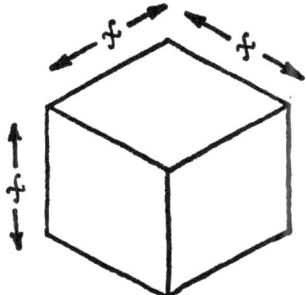

Figure 52

intended to represent merely the surface of the box, nor the three square faces that we can see in the picture.

With the help of the picture of x^2 we can perform simple multiplications in algebra. Suppose, for example, we wish to draw a picture of $(x+2)(x+3)$. As explained earlier in this chapter, we visualize this as $x+2$ rows with $x+3$ objects in each (Figure 53).

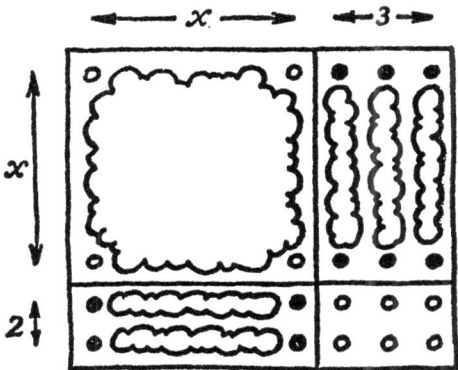

Figure 53

This breaks up naturally into several parts. In the north west, we see the square shape, x rows of x dots, that we have just learnt to associate with x^2. To the north east, we see 3 vertical columns, each containing x dots, and to the south west 2 horizontal rows, each containing x dots. Together, these contain $5x$

115

dots. Finally, we see 6 dots in the south east. Putting these pieces together, we obtain the result:

$$(x+2)(x+3)=x^2+5x+6$$

It is possible to carry out a number of interesting investigations, some theoretical and some practical, with pupils whose knowledge of algebra is confined to an understanding of pictures such as the one above.* It seems sensible to give children this experience of what can be done with a little algebra. As we have already seen, algebra involves a number of rather fiddling little distinctions, which it requires some care and effort to master. A pupil should surely be shown the kind of insight he can expect to gain if he perseveres with the subject. If he does not feel this type of insight to be worth having, he can the sooner make his decision to deploy his energies in some non-scientific and non-mathematical direction. If, on the other hand, he enjoys the samples of what algebra can achieve, this feeling of enjoyment will tend to spread to all his work in algebra, and will give him satisfaction in mastering some of the routine work which, to someone with no interest and sense of purpose, might appear tedious and dull. Teachers sometimes emphasize too much the element of boredom in study. 'You cannot hope to succeed in any subject,' they say, 'if you are not ready to submit to a certain amount of drudgery.' This is totally untrue. Things are not interesting or dull in themselves. They are interesting or dull according to how we feel about them. As R. L. Stevenson says:

It is in virtue of his own desires and curiosities that any man continues to exist with even patience, that he is charmed by the look of things and people, and that he wakens every morning with a renewed appetite for work and pleasure. Desire and curiosity are the two eyes through which he sees the world in the most enchanted colours: it is they that make women beautiful or fossils interesting.

What is true of every subject is that you cannot succeed unless you are willing to devote to it a considerable amount of time, thought, and activity. Some of this activity may be mechanical and pedestrian; this does not in any way mean that it need be

* See Chapter 8.

116

boring or dull. One's feeling about a subject tends to permeate all activities associated with that subject. Boredom arises only when a subject has been presented in such a way that it arouses neither curiosity nor desire. One's early experiences of a subject tend to have lifelong effects, and in nothing more than in the emotional tinge given to the work. It is, therefore, of the utmost importance that young children should first meet each subject in a way that makes clear that subject's interest, power, and purpose. When the desire to master the subject has been aroused, the technical details and incidental difficulties fall into their proper perspective.

CHAPTER SIX

Sudden Appearance of a Practical Result

THERE are several aspects of mathematics that arouse interest in children. To a young child, of course, everything is interesting, and so long as this state of poetic innocence has not been destroyed, mathematics will be interesting simply because it is something. There are other, more specific causes of interest. Some children like mathematics because it is orderly, reliable, and predictable. Much of adult life is, from a child's point of view – and indeed from that of any detached and thoughtful observer – wildly irrational and incomprehensible. But there are some areas of life in which adults and children meet on a more or less equal footing. Music and arithmetic are two of these. Musical ability can often be clearly detected in a child one year old, and mathematical ability not much later. These are subjects which children enjoy, because they belong to the world of children; children can master them. Arithmetic is one of the few subjects in which perfection can be attained. It is hard to imagine a picture, a piece of music, an essay, or a building that could not conceivably be surpassed. But in arithmetic a mark of 100 per cent can be given and can clearly be merited – and with proper teaching lies within the reach of almost every child. The element of surprise appeals to nearly all children. The 'think of a number' tricks have an almost universal appeal. There are surprises such as $3 \times 142857 = 428571$, where multiplication by 3 has switched the digit 1 from the beginning to the end, as was mentioned in Chapter 2. There are quieter surprises such as that 3×37 should be 111, written entirely with ones. There may be incidental pleasures. I remember a geometry book I had when I was quite young which gave instructions for drawing a spiral with the help of a pair of compasses. This gave me great satisfaction, not so much because of the theorem it illustrated, but because I enjoyed looking at the spiral and feeling I had drawn it. Then again, there is the aspect of mathematics which deals with the actual world. In children's magazines one sometimes sees little articles of this kind – 'How to find the height of a tree from the length of its shadow', 'How did the Boy Scout find the width of the river?' and so forth. These have a delusive appearance of being practical, but that is not why children are interested in them. They calculate the height of a tree because they like knowing how high the tree is, or because they enjoy measuring the shadow; they rarely have any engineering designs on the tree itself. Essentially these practical calculations are a form of play. Of course, like all play, they are a preparation for life, and give to everyone – not only to the future engineer – a keener appreciation of the objects that surround us.

Most children, in their receptive years between the ages of five and twelve, experience in some measure all these types of interest, from engineering to riddles and tricks. The strength of these interests varies from year to year and from child to child. We should try to teach in such a way that every kind of child is catered for. If we examine the earlier chapters of this book, we find the emphasis has been on tricks and riddles. It would indeed be very difficult to find any other effective way of presenting the beginnings of arithmetic and algebra. It is quite difficult to teach algebra to a boy who is, say, sixteen years old, and who insists that every remark made have a direct application to engineering. Algebra is indeed vital for engineering, but it is rarely that a chain consists of only one link. The first two or three links in the chain of algebra give little indication of the objects to which the far end of the chain is attached. It is for this reason that algebra is most easily taught to children eight or nine or ten years of age, whose interests are still general and in whom fantasy is lively. Even so, in our teaching we should include elements that will appeal to the realistic side of every child, and assure the strongly practical child that we have not forgotten him.

What we have done so far not merely fails to look practical; it looks as if it would never lead to anything practical. Consider, for example, the following two problems:

1. If you were in an aeroplane 1 mile above the surface of the earth, how far would you be able to see? We suppose, of course, that the day is clear, and that there are no mountains to complicate

things. The plane could be over the Atlantic or the plains of Illinois.

2. Two boys fix a telephone so that they can talk to each other when they are in bed. If their houses are 80 feet apart, and their bedrooms are 30 feet and 90 feet above the ground, how long a wire do they need?

It does not seem that we have learnt anything in this book that will help to solve either problem. Someone may say, 'I know, you are going to use a formula.' But we are not. A formula is indeed a useful thing. It puts in a concise form the fruit of some other person's experience; we accept it on trust in that person's competence and honesty. But in this chapter we shall quote no formula. We shall build only on what we have already developed, together with ideas that children already have. By this I mean not merely that children will accept the arguments I suggest, but that these arguments could be carried on in the form of a dialogue with children, the teacher merely asking questions. Naturally, the questions are chosen with some care. Indeed, they form a somewhat strange sequence, and at the outset it may seem unlikely that they will ever lead us to our desired goal.

AREA

One idea needs to be cleared up a little before we start, the idea of area. It seems to take children a long time to distinguish between *area* and the *distance round* some region.

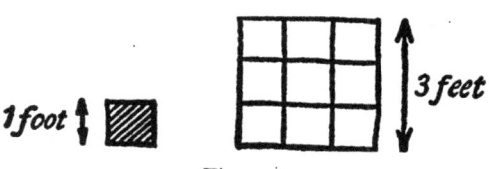

Figure 54

In Figure 54, children when asked for the area of the larger square will repeatedly answer, '12 feet'. This is, of course, the distance round the boundary; it is the length of the fence that would be required to enclose this region. When we ask for the area, we

Sudden Appearance of a Practical Result

want to know how many black squares would be needed to cover the square on the right. The answer, of course, is 9, which is 3×3 or 3^2. But time and again children seem to relapse into giving the distance round when the area is asked for. It seems necessary to emphasize this distinction, perhaps by having children actually pasting square pieces of paper over the region for which the area has been asked. Every time the word or the idea of area is involved, it is well to make some reference to pasting paper squares or to covering a floor, or to some similar illustration that will make clear which concept we have in mind. Anyone without classroom experience may be surprised how long it takes some children (including clever ones) to fix this distinction.

In the square 3 feet by 3 feet we see, of course, 3 rows of 3 squares each, so we have once more a picture of 3×3. If the square had sides 7 feet long, it would contain 7×7 squares.

The argument can also be made to work in the opposite direction. If a square can be covered by 9 black squares, it must be 3 feet by 3 feet. There is no other way of arranging 9 black squares so that they exactly cover a larger square. If 25 black squares exactly cover a larger square, that square must be of side 5 feet, for $5 \times 5 = 25$, and no other number will do.

Now we are ready to begin.

THE TILTED SQUARE

Suppose we have a square, *ABCD* (see Figure 55), each side of which is 7 feet long. We mark points *P, Q, R, S* on the sides of this square. *P* is 3 feet from *A*, *Q* is 3 feet from *B*, and so on, as shown in the figure. The distance from *P* to *B*, of course, is 4 feet, since the side of the square is 7 feet.

The picture we have drawn has a certain amount of symmetry. If we were to rotate it, so that *A* came to where *B* is now, *B* to where *C* is now, and so on, the appearance of the figure (apart from the lettering) would be unchanged.

What shape do the children think *PQRS* is? The general opinion seems to be that it is a square. We accept this verdict.

We now imagine that we have 49 square pieces of linoleum, each 1 foot by 1 foot. These would be just enough to cover the whole

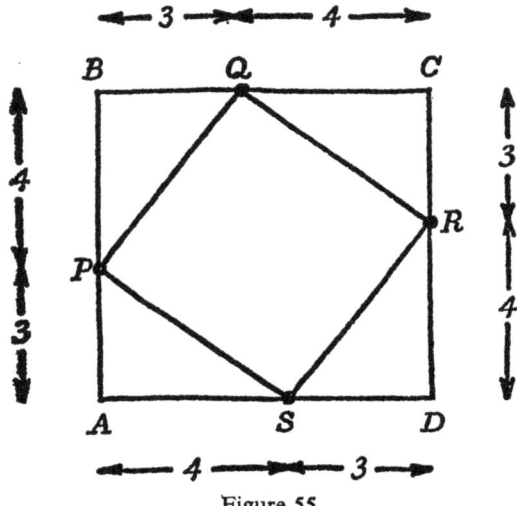

Figure 55

square *ABCD*. How many of them would be required to cover the tilted square, *PQRS*? We proceed to investigate this.

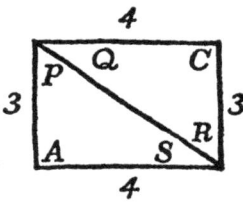

Figure 56

Suppose the triangle *PAS* in the south west and the triangle *QCR* in the north east were pushed together. They would form a rectangle, 3 feet by 4 feet, as shown in Figure 56. So, if we took 12 pieces of linoleum, arranged them in such a rectangle, and cut along the diagonal, we could cover the south-west and north-east corners in the square *ABCD*, which would now look as in Figure 57.

The subtraction at the side shows that we have used 12 of our 49 pieces of linoleum. We have 37 left to cover the unshaded area.

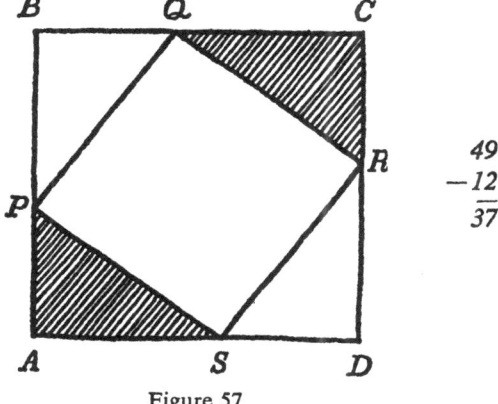

$$\begin{array}{r} 49 \\ -12 \\ \hline 37 \end{array}$$

Figure 57

What shall we do next? It seems reasonable to do the same thing with the north-west and the south-east triangles. These can be covered by using another 12 pieces.* The situation will then be as shown in Figure 58.

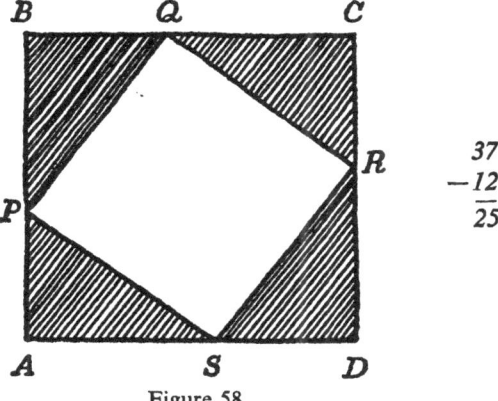

$$\begin{array}{r} 37 \\ -12 \\ \hline 25 \end{array}$$

Figure 58

As the subtraction shows, we are now left with 25 pieces on our hands. We know that we started off with just enough to cover

* The whole of this lesson will gain in vividness if it is demonstrated, with pieces of squared paper actually being pasted over the regions discussed.

the whole square *ABCD*. So presumably these 25 pieces are just enough to cover the still unshaded part, *PQRS*. But the only way that 25 pieces can be fitted together to make a single square is to have them in 5 rows of 5 (see Figure 59).

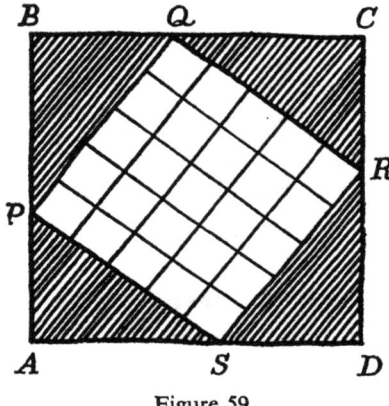

Figure 59

So we arrive at the conclusion that the distance from *P* to *S* must be 5 feet.

Everyone over a certain age has heard about the ancient Egyptians stretching ropes, of lengths 3, 4, and 5 units, in order to get the correct angle for the corner of a house, temple, or pyramid. This result has therefore lost some of its novelty. Even so, the method used above for arriving at this result may still appear somewhat remarkable. It is a very odd way of finding the length *PS* that we have used. Instead of discussing lengths, we have considered the areas of various regions, and this, in a way that would surely be totally unexpected to anyone who had not seen the method before, has led us to the length of the line *PS*.

The numbers 3, 4, 5 follow in order when we count, and some pupil will probably seize on this fact. Does this always happen? Could we get 7, 8, 9 for the lengths *PA*, *AS*, and *SP*? A few experiments will show that this is not so. If wooden bars, 7, 8, and 9 inches long, are bolted together, they form a triangle very different from those we have been considering. We fare no better

124

Sudden Appearance of a Practical Result

with 10, 11, 12, or 19, 20, 21. No, being in counting order is not the secret behind the numbers 3, 4, 5.

In our demonstration we began with a square 7 by 7, and broke each side into parts 3 and 4, in this way arriving at the triangle *PAS*. In a problem it is more likely that we should have to begin with the triangle. For example, we might be given the triangle, shown in Figure 60, with sides 5 and 12, and be asked to find the length of

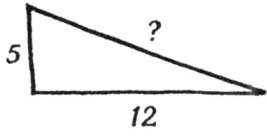

Figure 60

the third side. To solve this problem, we add 5 and 12. This gives 17, so we consider a square of side 17. Breaking each side into parts 5 and 12, we see the triangle we are interested in at the southwest corner of the square (see Figure 61). The pupils can be asked

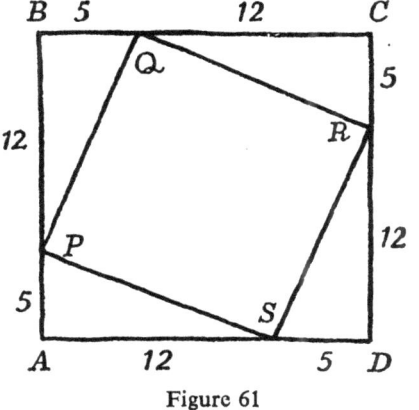

Figure 61

to carry through exactly the same argument with these new numbers. The arithmetic is a little more complicated now, but the strategy is the same. To cover the whole square *ABCD* would require $17 \times 17 = 289$ pieces. Of these 60 are required to cover triangles *PAS* and *QCR*, and another 60 to cover *PBQ* and *RDS*.

125

There remain $289-60-60=169$ to cover the tilted square *PQRS*. As $169=13\times 13$,* the length of *PS* must be 13.

Other numbers which give simple answers for *PS* are 6 and 8; 9 and 12; 12 and 16; 15 and 20; 7 and 24; 10 and 24. (The answers are respectively 10; 15; 20; 25; 25; 26.)

It will sometimes happen that the length *PS* is not exactly a whole number. For instance, if we make *PA* 4 inches long and *AS* 8 inches long, it can be checked by careful drawing that *PS* is about one twentieth of an inch less than 9 inches. This fact must appear in some way when we apply our method. We shall need to make *AB* 12 inches long. The area of the large square *ABCD* will be 144 square inches. Each pair of triangles will account for 32 square inches, so we shall arrive at $144-32-32$ $=80$ as the number of square inches in the tilted square. As $9\times 9=81$, we see that 80 pieces are just a little too small for making a square 9 by 9. It is reasonable that we could make a square of side just less than 9 inches out of these 80 pieces, though this poses a little puzzle – how, in fact, could you join 80 pieces together to make a square at all?

This type of puzzle arises for most of the triangles we can consider. For example, if we had started with 2 and 3 for *PA* and *AS*, we should have found the area of the tilted square to be 13 square inches. How can we take 13 inch-squares and join them to make a single square? If we had started with 1 and 1, the area of the tilted square would have turned out to be 2 square inches. How can we use 2 pieces, each an inch square, to make a single square? It is instructive to experiment with these problems before reading on.

AN ALTERNATIVE METHOD

In our work above, we found the area of the tilted square by a method involving subtraction; we considered a large square, and saw how much smaller than it the tilted square was. This, of course, was no help to us in solving the puzzles just mentioned; we can put pieces of squared paper together, thus illustrating addition of areas, but there is no process in solving a jigsaw puzzle

---

* This is easily guessed. $12\times 12=144$ is well known. We try something a little larger.

that suggests subtraction. It is, however, possible to build up the area of the tilted square by assembling its parts; this solves our puzzle and, at the same time, gives a way of checking results found by the other method.

The simplest example of all occurs when the numbers are 1 and 1, as in Figure 62.

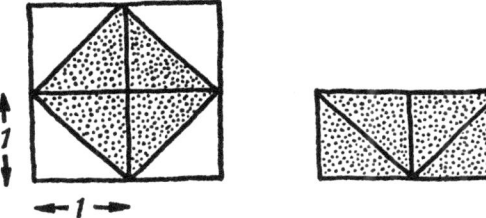

Figure 62

It is then fairly obvious that the tilted square, shown on the left, can be made from the two squares shown on the right.

When the numbers 1 and 3 are chosen, the situation is as shown in Figure 63. It is very natural to see the tilted square as made up

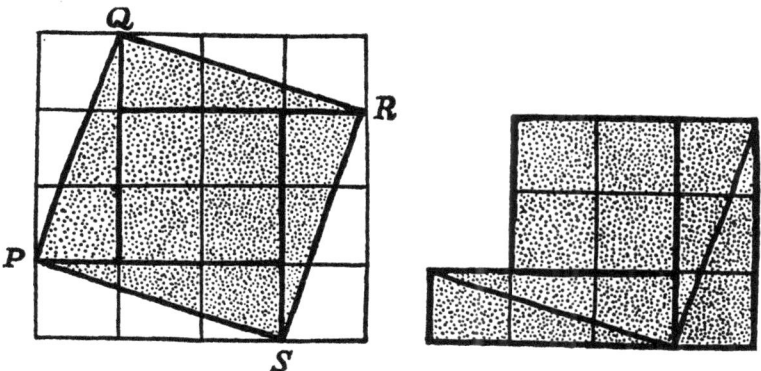

Figure 63

of the parts indicated. These can be rearranged so as to give the 10 squares shown on the right. A similar dissection provides the solutions to the other problems mentioned above.

In teaching this to children, one should remember that children often do not see shapes in the way that adults do. It may be perfectly clear to us that two triangles have the same shape and size: it may not be at all evident to a child. It is good to ask questions that will bring out the child's picture of the situation. For instance, one may mark on squared paper 2 points, such as *P* and *S* in our last illustration, and ask a child to indicate where *Q* and *R*, the other corners of this square, should be placed. One may draw some shape on a piece of squared paper, and ask the child to show how this shape would appear if it were moved to the right, or if it were turned through a quarter-turn. If a child gives strange answers to such questions, he should be given more opportunity to work with actual materials, so that he becomes familiar with the way shapes behave, and can visualize these correctly. It is no use trying to build on ideas that are not yet present in the child's mind.

THE UNDERLYING PATTERN

We now have two methods for finding the area of the tilted square. Both methods lead to the correct result, but they are both unnecessarily long. They can be replaced by a quicker method, which embodies a very well-known mathematical result. We now outline two methods by which pupils can themselves discover this result. Discovery is much more exciting than merely being told.

The first approach consists in making a table, showing many particular cases, and then trying to see some rule behind this table. We already have several cases to enter in this table; we have seen that when *PA* and *AS* are 3 and 4, the area of the tilted square is 25; we have seen that 4 and 8 give an area of 80; that 1 and 1 give an area of 2; 2 and 3 give 13; 1 and 3 give 10. These are entered in the table below:

PA	1	2	3	4	5	6	7	8	*PS*
1	2		10						
2			13						
3				25					
4							80		
5									
6									
7									
8									

128

Sudden Appearance of a Practical Result

The pupils should complete this table. One child's results may be checked against another's. The results may also be checked by using the two methods explained earlier – one using subtraction, the other addition – and seeing that they give the same answer. It is essential to have the correct numbers in this table, for if the numbers are wrong the pattern will be destroyed. It will then be impossible to discover the simple rule.

Completing the table involves work such as drawing squares or fitting pieces together which most children find enjoyable.

When the correct numbers are found the table shows evidence of great regularity. For example, the numbers in the top row are found to be 2, 5, 10, 17, 26, 37, 50, 65 which rise by steps of 3, 5, 7, 9, 11, 13, 15 – the odd numbers in order. The numbers in the second row are 5, 8, 13, 20, 29, 40, 53, 68 and these also rise by steps of 3, 5, 7, 9, 11, 13, 15. So do the numbers in all the other rows. Indeed, if anything, this table is too rich in patterns. The pupils may discover so many facts relating to *odd numbers* that they fail to realize that this table contains an even simpler pattern connected with *square numbers*. So it may be necessary to suggest that, as this problem is connected with squares – tilted squares, upright squares, inch squares – maybe the square numbers, 1, 4, 9, 16, 25, 36, 49, 64, have something to do with it. Each row should be taken separately and compared with the sequence of square numbers.

This completes our outline of the first way of looking for the pattern. We now come to a second approach. It is, incidentally, instructive to use both approaches; having solved the problem by the first approach does not destroy the interest of the second. This second approach does not use guessing as the first did. It simply translates into the language of algebra the method used in the arithmetic already done, and something becomes clear that was completely hidden in the arithmetic.

Our first calculation was based on the numbers 3 and 4, but, as we have seen since, exactly similar calculations could be made starting from any two numbers. So we try to separate the general procedure from the particular numbers. As a start, let us describe carefully the procedure used with the particular numbers 3 and 4.

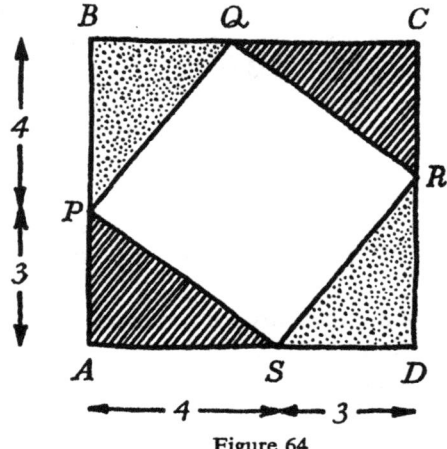

Figure 64

Step (i) We began by considering the square *ABCD* in Figure 64, of side 7. This number 7 is the sum of 3 and 4.

Step (ii) We found the area of the large square *ABCD*. This area is 49, which is 7×7, or 7^2.

Step (iii) We found the shaded area to be 12. 12 is 3×4.

Step (iv) We subtracted 12 from 49, which gave us the result 37.

Step (v) The dotted area was also found to be 12.

Step (vi) We again subtracted 12.

The position now is that from the whole area we have subtracted the shaded area and the dotted area. What remains is the area of the tilted square, which is what we are trying to calculate.

Having seen that this procedure works for *any* numbers, it is natural to consider what would happen if instead of 3 we used any number x that you cared to mention, and instead of 4 any number y chosen by me (see Figure 65).

We now go carefully through the various steps, replacing 3 by x and 4 by y.

Step (i) The side of the square is $x+y$.

Step (ii) The area of *ABCD* is the square of the number found in Step (i). This may be written as $(x+y)(x+y)$ or as $(x+y)^2$, in accordance with our language lessons in Chapter 5.

130

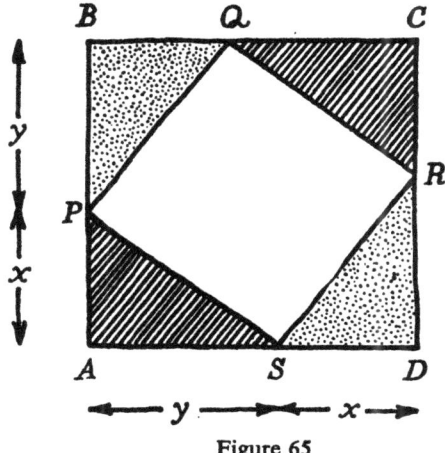

Figure 65

When we were dealing with 7×7, it was easy enough to work it out, and say it was 49. We shall have to spend a little more time working out $(x+y)^2$. In Chapter 5 we saw how to picture such things; $(x+y)^2$ is the number of objects present when you have $x+y$ rows, each row containing $x+y$ objects (Figure 66).

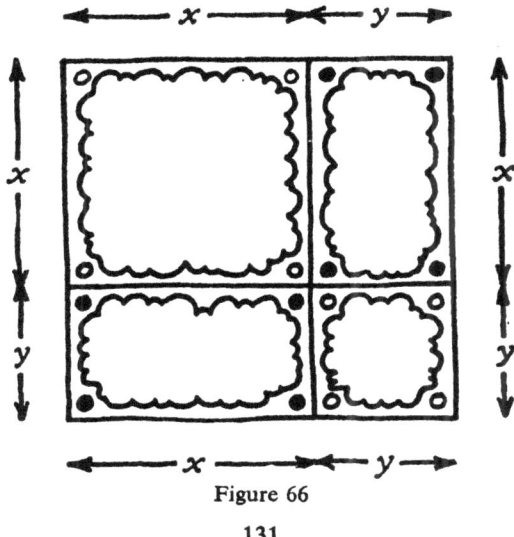

Figure 66

Vision in Elementary Mathematics

This picture breaks up into parts. We see a square containing x rows of x objects; this represents x^2. We see, at the bottom right-hand corner, a square with y rows of y dots, representing y^2. We see also 2 rectangles; one is standing up and one is lying on its side, but they are of the same size; each of them contains xy objects (if necessary, we remind the children that this means x times y). As there are 2 such rectangles, together they contain 2 xy objects; $2xy$ may be read as 'twice xy' to emphasize its meaning.

Altogether, then, the square contains x^2+y^2+2xy objects, and this is as far as we can go in working out $(x+y)^2$.

So we have:

$$(x+y)^2 = x^2+y^2+2xy.$$

This completes Step (ii). Now Step (iii) tells us to find the shaded area in Figure 64. In our particular case, the answer was 3×4. This strongly suggests the general answer x times y, since x replaces 3 and y replaces 4. And, in fact, this answer is correct, for the shaded area can be made from a rectangle x by y. The shaded area is thus xy.

Step (iv) tells us to take the answer in Step (iii) from the answer in Step (ii), that is, to subtract xy from x^2+y^2+2xy. If we subtract xy from twice xy the result is xy. So, Step (iv) leaves us with the result x^2+y^2+xy.

Step (v) merely notes that the dotted area is the same size as the shaded area, that is to say, xy.

Step (vi) tells us to subtract that also. So we subtract xy from x^2+y^2+xy, and obtain the final answer x^2+y^2.

So the area of the tilted square is x^2+y^2. This checks with our particular case; if you choose 3 for x and I choose 4 for y, then x^2+y^2 will become $3^2+4^2=9+16=25$, as we had before. In the same way, if we chose 1 for x and 3 for y we should find the answer $1^2+3^2=1+9=10$, the same result that we obtained earlier by arithmetic. Arithmetic gives these particular answers, 25 and 10, but it conceals the simple rule behind them. It does not show that 25 is the sum of the 2 square numbers 9 and 16, nor that 10 is the sum of the 2 square numbers 1 and 9. But in algebra,

Sudden Appearance of a Practical Result

x^2+y^2 shows clearly that the answer is found by adding together two square numbers, x^2 and y^2.

Now x^2 is something we can picture; it gives the area of a square of side x. In Figure 65, we already have a line of length x, namely PA. So we can picture x^2 by means of a square with side PA. In the same way, we can picture y^2 as giving the area of a square with side AS. If we draw these squares on Figure 65, and omit the points B, C, D which were useful in the proof but serve no purpose now, we arrive at the picture in Figure 67.

The area of the tilted square $PQRS$, being x^2+y^2, is equal to the combined area of the 2 shaded squares. That is to say, if these 2 squares were covered with material, such as cloth or cardboard, it would be possible to cut this material up, and reassemble the pieces, so that they just covered the square $PQRS$. Here we have another jigsaw puzzle. How to solve it may be suggested by Figure 63 on page 127.

If we use the letter z to indicate the length* of PS, the area of the square $PQRS$ will be z^2. The shaded area is x^2+y^2. Since the area of the square $PQRS$ is the same as the shaded area, we have the result:

$$x^2+y^2=z^2$$

This result allows us to shorten our calculations considerably We no longer need to use the process described in Steps (i) to (vi). To return to our first example of all, where the sides are 3 and 4. As $x=3$, $x^2=3\times3=9$. As $y=4$, $y^2=4\times4=16$. Then $z^2=9+16=25$, so z must be 5. Our result is, of course, an old and very familiar one, the Theorem of Pythagoras, known to Gilbert and Sullivan's Major General who was:

'teeming with a lot o' news –
With many cheerful facts about the square on the hypotenuse.'

---

* Since z represents a number, the length of PS will be z feet or z inches or z miles or z centimetres, according to what units are being employed. Thus, strictly z does not represent the *length* but the *number* of units in the length. It is tedious to repeat this every time a symbol is introduced. For brevity, we speak of a line 'of length x' or a square 'of area x^2'. The reader may decide for himself in what units he will think. If he decides on inches, then the length intended above is x inches, the area is x^2 square inches.

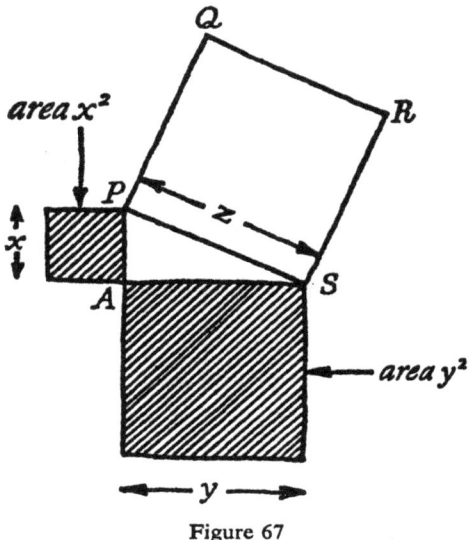

Figure 67

THE ROAD TO THE RESULT

The subject of this chapter illustrates a remark made earlier, that non-mathematicians tend to be too humble. They feel that, to an intelligent person, the results of mathematics should be obvious. But the road to Pythagoras' Theorem is anything but obvious. A traveller in ancient Greece reports that the Egyptians use an interesting device for constructing right angles – they make a triangle with sides 3, 4, and 5. It is natural to wonder about the significance of these numbers. As we have seen, the fact that 3, 4, 5 follow in order when we count is not at all relevant. What then is the important property of these numbers? It seems to me that one might rack one's brains for hours, without ever guessing the key – that 3 threes and 4 fours together make 5 fives. The problem does not yield to direct attack. It may well be that the solution was found by someone who was thinking about a different problem altogether. Perhaps he had tried to explain the Egyptian method and given it up as a bad job. Then one day he was looking idly at a tiled floor, and began to think about the number of tiles needed for a tilted square such as that shown in Figure 62. He might next think about a larger square, such as the one in Figure 68.

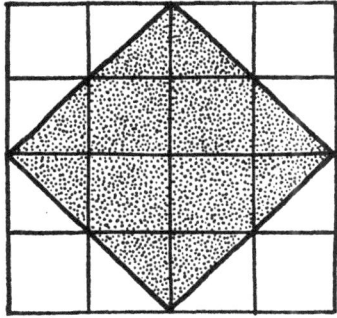

Figure 68

By counting whole tiles and half tiles he would easily arrive at the area of this square. Still regarding this activity simply as an amusing puzzle, he considers a square such as that in Figure 63, involving the numbers 1 and 3. He sees that he can find the area of any tilted square, and, among other results, finds that the numbers 3 and 4 lead to the answer 25. At some stage of the work, associations begin to stir in his mind; 3, 4, 5, the Egyptian triangle; 5 fives are 25 – and suddenly the flash comes with great intellectual excitement: why, here is the explanation of the Egyptian method. The hardest part of the discovery has been made; the way to find the length of *PS* is to consider the area of the square *PQRS*. But this is not yet the end of the road. 5 fives have come into the picture as the area of the tilted square, but there is as yet no hint that 3×3 and 4×4 are also involved. The next step, perhaps, was to complete a table such as that on page 128. As we have seen, the connexion of the numbers in this table with the square numbers is not too obvious, but it can be discovered. So in due course the law $x^2 + y^2 = z^2$ is guessed as one that fits this table. It works so long as x and y are whole numbers not larger than 8. But it still has to be proved that this law works whatever numbers x and y may represent. That might be achieved by using the type of dissection shown in Figure 63.

135

Vision in Elementary Mathematics

The account we have just given of the way in which some early mathematician might have arrived at Pythagoras' Theorem, resembles the attack outlined in 'An Alternative Method' on page 126. Why, if this is the natural approach, have we presented it as the alternative and not as the main method? The reason is that we are dealing with pupils who are supposed to know only a small amount of algebra. Our method requires the pupil to express $(x+y)^2$ in the form x^2+y^2+2xy, and to subtract xy twice, leading to the answer x^2+y^2. If we had used the dissection shown in Figure 63, this would have been more natural, and we would have used addition rather than subtraction. However, when we came to express our result in algebra, we should have been led to the expression $(y-x)^2+2xy$. This is indeed equal to x^2+y^2 but, as we have not yet discussed $(y-x)^2$, the pupils would not have been able to deal with this expression.

These considerations illustrate the way a mathematician probes his way into a problem. He does not just have one method and stick to it through thick and thin. He weighs up several ways of attacking a problem. One may seem promising at the beginning but run into difficulties later on. The mathematician then returns to his starting-point and tries another path, until he finds one that takes him all the way – or perhaps abandons the problem in despair.

The advantage of a little extra knowledge is also indicated. If a person knows how to deal with both $(x+y)^2$ and $(y-x)^2$, it does not matter which method he chooses. Either way, he will be able to force through to the end. But someone whose knowledge is limited has to use great foresight in his choice of method if he is not to be blocked by meeting calculations beyond his powers.

THE BOYS' TELEPHONE

Now at last we are able to deal with the problem of the boys' telephone. In Figure 69, A and B are the ends of the wire at the boys' bedrooms. It is helpful to show the level line BC, for the triangle ACB is now of the kind we have been discussing throughout this chapter.* If we can find x and y, Pythagoras' Theorem

---

* Mathematicians call such triangles *right-angled*.

136

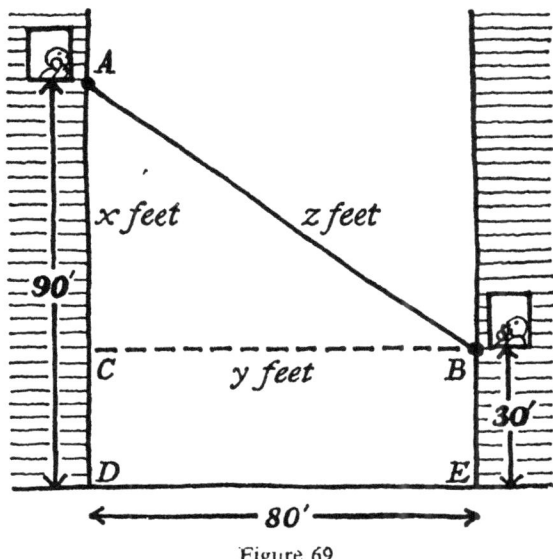

Figure 69

will show us how to find z. Now it is not at all hard to find x and y. *BCDE* has the shape of a brick (a rectangle), and the top of a brick is just as long as the bottom. So *CB* and *DE* must be the same length. This means that y is 80. Again, the height of a brick is the same at both ends. So *CD* must be the same length as *BE*, that is, 30 feet. This gives us the length of *AC*, for *AC* is what remains when you cut *CD* off *AD*. 'Cutting off' means subtraction, so *AC* is 60 feet, and x is 60.

We now use $z^2=x^2+y^2$ to find z. As $x=60$, $x^2=60\times60$ $=3,600$. As $y=80$, $y^2=80\times80=6,400$. So $z^2=x^2+y^2=3,600$ $+6,400=10,000$. As $100\times100=10,000$, $z=100$. So the wire *AB* must be 100 feet long.

The method used in the telephone problem is of a very wide application. The essential point is that we go from A to B via C. The journey from A to C is vertical: from B to C is horizontal. Any journey whatever can be broken in this way, into one part that is up or down, and a second part that is across. This idea is the key to the use of squared paper, or of a grid in maps. For

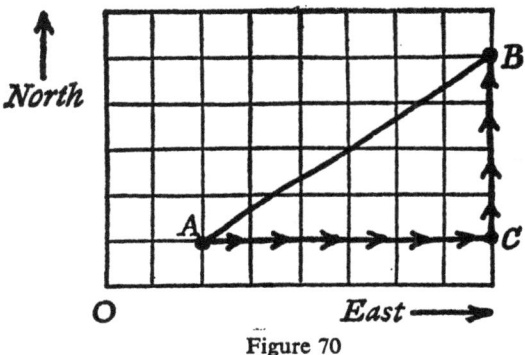

Figure 70

example, in Figure 70, A and B are 2 towns. Everything is measured from the town O. The position of A is specified by its being 2 miles east and 1 mile north of O. Town B is 8 miles east and 5 miles north of O. How far is it from A to B in a direct line? If we travel from A to B via C, our journey consists of 6 miles to the east followed by 4 miles to the north. As $6 \times 6 = 36$ and $4 \times 4 = 16$, if AB is z miles, $z^2 = 36 + 16 = 52$. Now 52 is a little more than 49, or 7×7. So z is a little more than 7, and AB is a little more than 7 miles. If a more accurate value is required, it can be found from a table. In a table of *square roots* one will find entries such as these:

48	6·9282
49	7
50	7·0711
51	7·1414
52	7·2111
53	7·2801

From this table we see that if $z^2 = 52$, then (to 4 places of decimals) $z = 7 \cdot 2111$. This answer is, in fact, too accurate! The fourth place of decimals represents 1/10,000. Now 1/10,000 mile is about 6 inches. It is meaningless to give the distance between 2 towns to the nearest 6 inches.

Children should be allowed to use tables of square roots in such work, for this allows them to think about the method without becoming lost in the details of the calculation. Using the tables will help to make the pupils familiar with the meaning of square root. Before looking up a particular result, they should be encouraged to guess how large the answer should be. For example, before looking up the square root of 41, the pupils should recall that 41 comes between 36 and 49, so we expect a number between 6 and 7. The table gives the value 6·4031, which fulfils our expectations.

In Figure 70 we showed the towns A and B on squared paper, and the position of each town was given by means of two numbers, representing distances east and north from O. This idea, of representing positions by numbers, is frequently useful. It can be used to record the motion of a planet round the sun, of an object thrown into the air, of a satellite around the earth, of a cog in a machine, and for many other purposes. The development of this idea leads to the mathematical subject known as *coordinate geometry* (i.e. geometry through numbers). The central result of coordinate geometry we have already met in this chapter – the theorem of Pythagoras, applied in the way we used on page 138 to find the distance between points A and B. From this, many results of coordinate geometry can be developed in a very simple and natural way. These developments will not be discussed further here, since they require the pupil to be quite familiar with algebra and geometry. We merely note in passing that here is a seed from which a large tree can grow.

THE DISTANCE TO THE HORIZON

Suppose a person is poised above the earth, in a balloon, an aeroplane, a satellite, or even at the top of a mountain, a skyscraper, or a high tower. What determines how far he can see? In Figure 71 the person at P certainly cannot see an object at A on the other side of the world. He still cannot see the object if it moves to B or C. As the dotted lines show, we could only see from P to C if our vision could pierce through the solid earth. The object will just come into sight when it reaches the position D. The distance we want to find is PD; this is the distance from the person P to an object D, which is as far away (on the surface of the earth) as P can see. We regard the earth as being perfectly

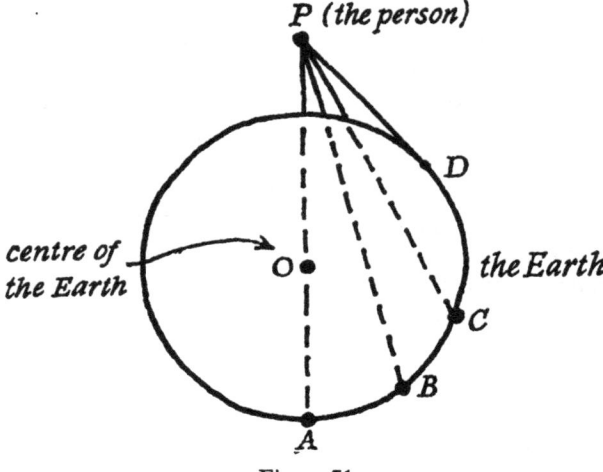

Figure 71

round, and of radius 4,000 miles. So OD is a distance of 4,000 miles. Suppose P is in a plane 1 mile above the surface of the earth. Then OP is 4,001 miles. These facts are collected in Figure 72. Now we ask our pupils what they think about the triangle

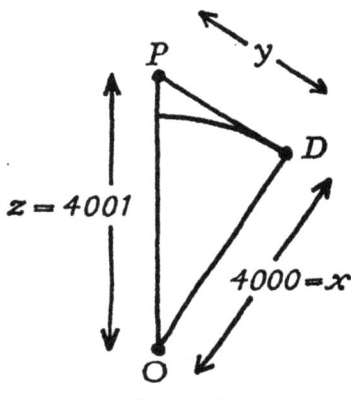

Figure 72

ODP. Do they notice anything about it? Or is it just any old triangle? Most children will say that it *looks* as if the triangle had

140

Sudden Appearance of a Practical Result

a right angle at D. That is indeed the case. In formal geometry one can give rather more elaborate arguments in support of this conclusion. We do not need to go into these here. The angle at D looks like a right angle: the children think it is; we accept their verdict. Now we are in a position to apply the Theorem of Pythagoras, $x^2+y^2 = z^2$. On this occasion we know x and z; Pythagoras' result will allow us to find y. Using the facts that $x=4,000$ (so $x^2 = 16,000,000$) and $z = 4,001$ (so $z^2 = 16,008,001$), we get the formidable looking equation:

$$16,000,000+y^2 = 16,008,001$$

At this point, one should be prepared for the children finding a small difficulty, because the equation contains y^2. If the equation had been $16,000,000+k=16,008,001$, they would have had no difficulty in saying 'k must be 8,001'. But y^2 looks more complicated. They do not see so easily that y^2 stands for a single number. You may find it helpful to cover y^2 with your hand, so that the pupils see $16,000,000 + \ldots = 16,008,001$. The dots represent what your hand covers. You now ask what number must be covered by your hand. The children say, '8,001.' Could it be anything else? No: nothing else will do; the number under your hand must be 8,001. Then you move your hand, and y^2 is seen. So y^2 must be 8,001. As $90 \times 90 = 8,100$ and $89 \times 89 = 7,921$, we see that y lies between 89 and 90. Tables give the more precise value 89·45.

By exactly the same method, we can calculate the distance that can be seen by a person at any height above the earth's surface.

Exercises

1. What would be the distance PD:
    (i)   for a person 2 miles above the earth?
    (ii)   for a person at a height of 3 miles?
    (iii) for a person at a height of 4 miles?
    (iv) for a man in orbit at a height of 100 miles?

2. Figure 73 represents paper ruled in inch squares. What are the distances:

    (i) A to D?
    (ii) B to C?

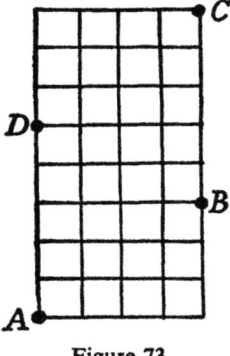

Figure 73

(iii) *A* to *B*?

(iv) *D* to *C*?

(i) and (ii) are obvious; (iii) and (iv) can be found by the method on page 138.

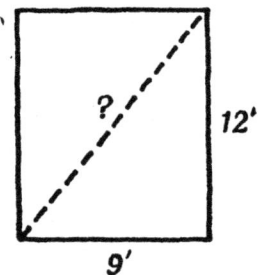

Figure 74

3. A room is 9 feet wide and 12 feet long (see Figure 74). In order to check whether the angles are correct, a man measures the distance between opposite corners. What should this distance be? (This method can be used to check the layout of a tennis court, or any other object in the form of a rectangle.)

4. The 2 lines in Figure 75, of lengths s and t, look about the same length. Find s^2 and t^2 by the method of page 138 and so decide whether the lines are exactly the same length or not.

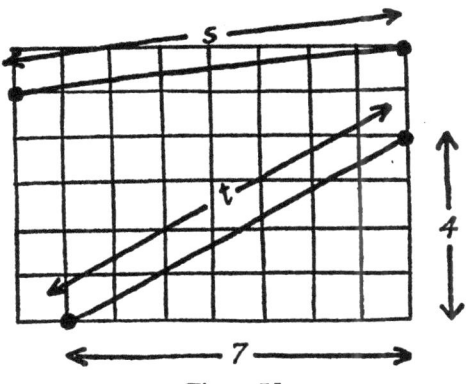

Figure 75

Figure 76

5. A carpenter intends to draw a line *AB* 10 inches from the edge *CD* of a piece of wood (see Figure 76). However, he is an exceptionally bad workman, and instead of having the pencil point *A* directly above his hand, he has it 1 inch to the right, as shown. What will be the actual distance, *y* inches, between the line *AB* and the edge *CD*?

(The square root of 99 is very close indeed to 9·95.)

143

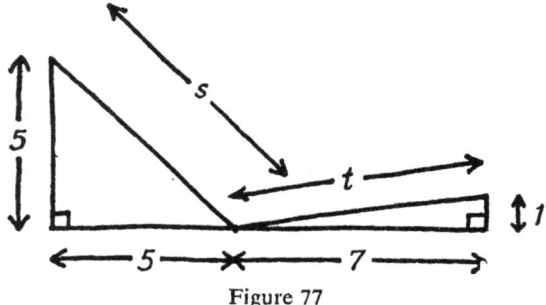

Figure 77

6. Are *s* and *t* equal or unequal (see Figure 77)?

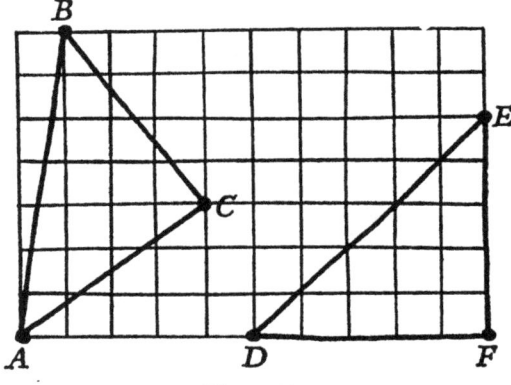

Figure 78

7. In Figure 78 compare (i) the distance from *A* to *C* with the distance from *D* to *F*, (ii) the distance from *C* to *B* with the distance from *F* to *E*, (iii) the distance from *A* to *B* with the distance from *D* to *E*. (The method of page 138 may be used for those distances which are not obvious.)

Would it be possible to cut out the triangle *DEF* and place it so that it just covered the triangle *ABC*?

CHAPTER SEVEN

A Miniature Problem in Design

WHEN some complicated contrivance is being designed, each decision involves a long chain of consequences. If we alter the size or the weight or the strength of one part, we may have to re-design every other part. Often it is hard to see all the effects of each decision; we would like to postpone the moment when we commit ourselves. Algebra gives us a way of doing this. Instead of choosing any particular length, we label some part as being x feet long. We work out the sizes of all the other pieces, in terms of x. This shows us how the choice of x will affect them. Only after doing this do we replace x by a particular number.

It is difficult to illustrate this procedure by examples drawn from actual engineering or scientific practice. The calculations are usually too involved, the technical considerations too many, for the work to serve as an introduction to elementary algebra. One is, therefore, forced to consider somewhat unreal situations. The problem discussed in this chapter is far from the normal routine of engineering. It is about a man who is troubled by bears, and builds a series of gates to keep them out. This problem does, however, illustrate the essential feature: the sizes of the gates are all interdependent; fixing the size of one gate fixes the sizes of all. Further, the solution of the problem involves precisely one principle in algebra. As was mentioned earlier, pupils often learn to simplify expressions such as $50-(20-x)$ without feeling any compelling urge to study such things. The present chapter provides one illustration of how the need for such manipulations can arise.

When we were explaining tricks in Chapter 4, we used a bag to represent the number thought of. This device showed that a long chain of instructions could often be replaced by something much shorter. For example, the final effect of the instructions 'Think of a number; add 2; double; add 1; multiply by 3' is represented by 6 bags and 15 stones. The instructions are, therefore, equivalent to: 'Think of a number; multiply by 6; add 15.'

In the bear-and-gate problem, we shall need to find a shorter equivalent for the following type of instructions: 'Think of a number. Subtract it from 20. Subtract the result from 50.'

This is not easy to illustrate by means of bags. We can investigate it by supposing different children to think of different numbers.

	Alf	Betty	Charles	Dora	Ernest
Think of a number	1	2	4	7	9
Subtract from 20	19	18	16	13	11
Subtract result from 50	31	32	34	37	39

It is readily seen that there is a quick way to pass from the numbers in the first row to those in the third, namely to add 30 to the number thought of. Starting with x, we end with $x+30$. The instructions as originally given require us to start with x, then find $(20-x)$, and finally to subtract this from 50, to obtain $50-(20-x)$. Brackets are necessary here as was explained in Chapter 5. It seems then that $30+x$ is a shorter way of saying the same thing as $50-(20-x)$. A pupil working out $50-(20-x)$ is tempted simply to ignore the brackets and write $50-20-x$. This, however, is incorrect. It begins all right; $50-20$ is 30. The ending, however, is wrong. The answer should not be $30-x$ but $30+x$. The correct scheme is:

$$50-(20-x)$$
$$=50-20+x$$

One can see that this result is reasonable. Suppose I have £50 in my pocket and I go to buy a suit costing £20. The tailor, however, is in a sporting mood; he says he will reduce his price by a number of pounds, to be fixed by drawing a number out of a hat. If x stands for the number so drawn, I shall not have to pay 20 pounds but only $(20-x)$ pounds. Instead of returning home with $50-20$ pounds, I shall return with $50-(20-x)$ pounds. Now clearly the more he reduces the price by, the better off I am. I hope x will turn out to be as large as possible, for I bring home $30+x$ pounds. If I were going to bring home $30-x$ pounds, I would hope that x would be small – and this obviously does not fit the story. We might illustrate the position as in Figure 79.

146

A Miniature Problem in Design

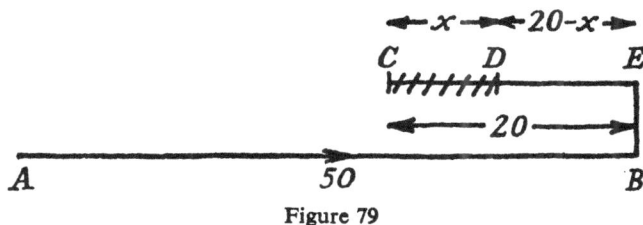

Figure 79

The point A represents the situation in which I have nothing. The story unfolds; 'I have £50'; this brings me to B. I am expecting to pay out 20, so I expect to go back to C. The reduction in price is shown by the crossed-out line, CD. Instead of having to go back from E to C, representing a payment of 20, I only have to go back from E to D, representing a payment of $20-x$. So D represents $50-(20-x)$. The history of my hopes and fears could equally well be shown as in Figure 80. At B, I have 50; a payment

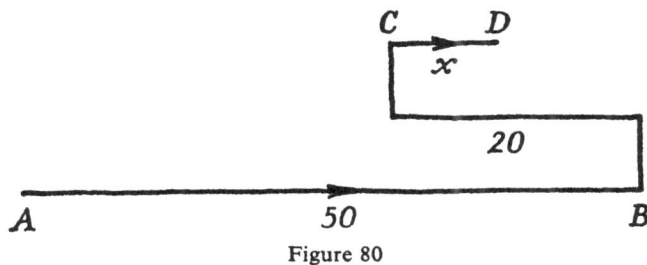

Figure 80

of 20 would take me back to C; a refund of x brings me once more to D. A movement to the right represents something added to my fortune, to the left, something taken away. Figure 80 thus illustrates the meaning of $50-20+x$. The final position, D, is the same as we had in Figure 79 to illustrate $50-(20-x)$.

This principle is not a particularly difficult one, when taken by itself. The trouble, of course, is that algebra involves a mass of similar small ideas, and pupils very easily get them mixed up. It is, therefore, essential that children should have plenty of time to meditate on each idea, and develop some way of visualizing it or thinking it out. Some children may be helped by experiments with particular numbers. What signs should we put in 10 ... 4 ... 3 to make it equal to $10-(4-3)$? The pupil, of course, must be clear that $10-(4-3)$ means $10-1$, which is 9. It will be found that only $10-4+3$ gives this result. The pupils may make up stories to illustrate the meaning of an expression such as $10-(4-3)$.

A SUBTRACTING MACHINE

The following device shows an interesting aspect of subtraction. A long strip of paper is laid on a ruler. The first few numbers, 0, 1, 2, 3, 4, are marked on the paper strip, which is then folded over at, say, the point where 10 would be marked (see Figure 81). It

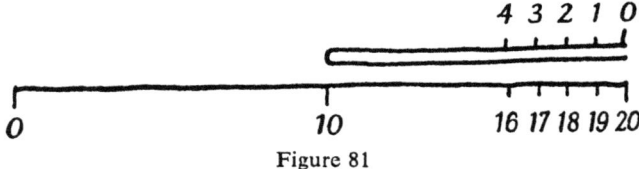

Figure 81

will be found that 0 then lies over the mark 20 on the ruler, 1 lies over 19, 2 over 18, and so on. In fact, x lies over $20-x$; this is a machine for subtracting from 20. It will be seen that subtraction produces a reversal; the numbers 0, 1, 2, 3, 4 originally ran from left to right; they now run from right to left.

To subtract from 20 we folded the strip at the mark 10, which is half of 20. If we want to perform another subtraction, we shall have to fold our tape again. To subtract from 50, we must fold at the mark 25, and reach the situation shown in Figure 82.

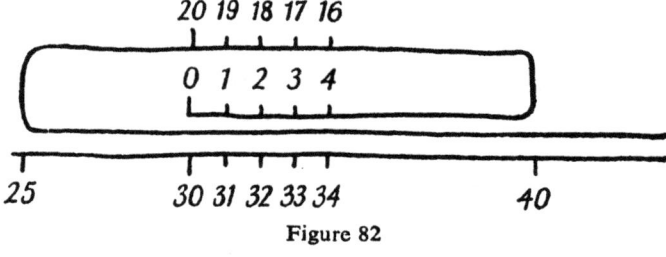

Figure 82

148

A Miniature Problem in Design

Above 2 we see 18, the result of subtracting 1 from 20. Below 2, we see 32, the result of subtracting 18 from 50. It is the same for the other numbers. Above x we see $20-x$; below it we see $50-(20-x)$. It will be noticed that the numbers 0, 1, 2, 3, 4 again read from left to right, as they did originally. Each comes over a number exactly 30 more than itself. In fact, *so far as this part of the strip is concerned*, the effect is exactly as if it had been allowed to slide 30 units to the right – an operation that we associate with addition.

The danger of verbal learning is shown here. The effect of one subtraction followed by another has, *in this situation*, been equivalent to an addition. Subtract from 20; subtract the result from 50; the combined effect is to add 30. But a pupil who parrots '2 subtractions make an addition' or '2 minuses make a plus' or some other slogan of this kind, is liable to think that $100-2-3$ should be 105, for are there not 2 minus signs here? Yet it is surely evident that if you have 100 objects, remove 2, and then remove 3 more, you certainly do not end with 5 more than 100. It is probably desirable for children to discuss their mathematical experiences and to formulate their conclusions in rules and to have some sort of verbal labels – 'Pythagoras' Theorem'; 'the difference of two squares'; 'you can add numbers in any order'. The real difficulty of mathematics comes in deciding which label applies to a given situation. The pupil must have his own way of remembering, seeing, and thinking this out. A hastily memorized rule is entirely insufficient.

Exercises

1. What signs are needed to complete the following statements?

$$9-(7-3)= \quad 9 \quad 7 \quad 3$$
$$9-(7-2)= \quad 9 \quad 7 \quad 2$$
$$9-(7-x)= \quad 9 \quad 7 \quad x$$
$$10-(6-5)=10 \quad 6 \quad 5$$
$$10-(6-1)=10 \quad 6 \quad 1$$
$$10-(6-x)=10 \quad 6 \quad x$$

2. Think of a number; subtract it from 10; subtract the result from 11. What is the final effect of these instructions? Experiment by thinking of different numbers.

149

Questions (3) – (8) are variations on question (2). In each question instructions are given, and we want to know the effect of these on any number anyone mày think of. The questions are entirely separate.

3. Subtract from 10; subtract the result from 12.

4. Subtract from 15; subtract the result from 25.

5. Subtract from 20; subtract the result from 20.

6. Subtract from 10; subtract the result from 20; subtract the result of that from 30.

7. Subtract from 10; subtract the result from 30; subtract the result of that from 100.

8. Subtract from 10; subtract the result from 20; subtract the result from 30; subtract the result from 25.

9. Which of the questions above is connected with the expression $25 - (15 - x)$? What does it show this expression to be equal to?

10. Which question is connected with $12 - (10 - x)$? What is the simplest form of this expression?

11. Which question is connected with $20 - (20 - x)$, and what is the simplest form of this expression?

12. Questions (6) and (7) involved 3 subtractions. Question (8) involved 4. One could make up questions with even more subtractions. Experiment with these.

What is the final effect of such a procedure with (i) an *even* number of subtractions, (ii) an *odd* number?

THE BEARS PROBLEM

We now have the algebra necessary for our problem. In Figure 83 there are 3 posts, *A*, *B*, and *C*. The distances between them are as shown. The man wants to mount *one* gate on each post. The gates are shown by *AL*, *BM*, and *CN*. As they are shown in Figure 83, these gates do not keep anything out. But, if bears appear from the north west, the man will swing the gate *BM* into the position *BL*, and the gap between the posts *A* and *B* will be closed. If the bears come from the north east, the man will leave gate *BM* where it is, and swing gate *CN* to position *CM*, so as to close the side *BC*. If the bears are coming from the south, the man will have 2 of his gates in positions *AN* and *NC*, and thus close

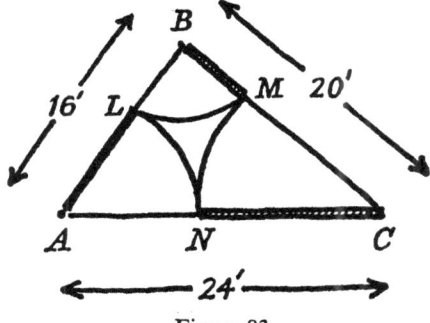

Figure 83

side *AC*. The problem is: if all the gates are to meet neatly without overlap, how big should each be?

Now, of course, this problem can be solved quite easily by trial and error. Suppose we first try 3 feet for the size of the gate at *A*, as in Figure 84. To close the side *AB*, the gate at *B* will have to

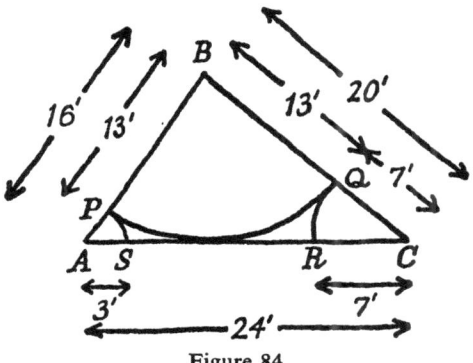

Figure 84

be 13 feet. This gate swings round to the position *BQ* when it is needed for side *BC*. The gate at *C* has to be 7 feet. But now we see that this is no solution. When the bears approach from the south, we have gates of 3 feet at *A* and 7 feet at *C*, leaving a gap of 14 feet in the middle, which the bears would pour through.

So we try again. If we enlarge the gate at *A* to 4 feet, that at *B* will be 12 feet, and at *C*, 8 feet, if the sides *AB* and *BC* are to close

151

properly. For the 24 feet of *AC*, we now have gates of 4 feet and 8 feet; the gap has been reduced to 12 feet. We are changing in the right direction. One might tabulate further guesses in this way:

Gate at *A*	3	4	5	6	7	8	9	10
Gate at *B*	13	12	11	10	9	8	7	6
Gate at *C*	7	8	9	10	11	12	13	14
Gap *SR*	14	12	10	8	6	4	2	0

The last guess makes the gap disappear, and gives us the solution. Making such a table is instructive, and for many children this will be the simplest and most convincing way of solving this problem. However, further lessons can be drawn from the table above. This table has a connexion with the repeated subtractions earlier in this chapter. We choose any number for the top row, *Gate at A*. This is like the instruction, 'Think of a number'. To obtain the number in the second row, *Gate at B*, we subtract from 16. This is natural, since the gates at *A* and *B* have to fill the 16 feet of the side *AB*. We also do a subtraction to get from the second to the third row: we subtract from 20, because the gates at *B* and *C* have to fill 20 feet.

Thus, if the number in the first row is indicated by x, that in the second row is $16-x$, and in the third row we have $20-(16-x)$. This, as we saw earlier, is the same as $20-16+x$, that is $4+x$. In the table it will be seen that the numbers for *Gate at C* are, in fact, 4 more than those for *Gate at A* (see Figure 85).

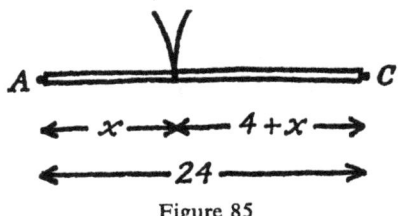

Figure 85

Now we want to choose x in such a way that the gate at *A* and the gate at *C* will be just big enough to fill the side *AC*, which is 24 feet long. This is an addition situation. We hope that x added to $4+x$ will make 24 (see Figure 85). The addition of x and $4+x$

152

A Miniature Problem in Design

is easy – a bag added to a bag and 4 stones gives 2 bags and 4 stones. The sum is $2x+4$. So we want to choose x in such a way that:

$$2x+4=24$$

Clearly we should choose 10 for x, which, of course, agrees with the solution found earlier by trial and error.

Children sometimes cling to the trial-and-error method because, although it involves more work, it does not involve any new idea. It is no use forcing them to use algebra, by saying that this question must be done with x. If the idea of x is introduced to children early and skilfully, they will become familiar with x and proud of being able to use it. They will see that some problems can be solved much more quickly by algebra than by arithmetic. Examples of such problems will be found among the exercises given later in this chapter.

It may be useful to collect together the considerations we have used in the Bears Problem, and present these in the form they might have in an exercise worked by a pupil – without the asides a teacher uses in presenting the method. A solution might read as follows:

'Let Gate A be x feet in length. Then Gate B must be $16-x$ feet, to make gates A and B fill side AB. Then Gate C must be $20-(16-x)$ feet, to make gates B and C fill side BC. Hence, Gate C is $4+x$ feet long. Gates A and C together should fill 24 feet. So $2x+4=24$. Therefore, $x=10$. Hence, $16-x=6$ and $4+x=14$. Gates A, B, C, should be 10, 6, 14 feet respectively.'

Exercises

In the problem discussed above, the distances between the posts were 16, 20, and 24 feet. An obvious way to make exercises is to replace 16, 20, and 24 by other numbers. What sizes would the gates have to be if the distances between the posts had been:

1. 16, 18, 20 feet
2. 30, 40, 50 feet
3. 3, 4, 5 feet
4. 20, 80, 90 feet
5. 11, 12, 21 feet

In each case, make a sketch to show the solution and check that it really works.

ANOTHER SMALL RESULT

My sporting tailor might have suggested that, instead of reducing the cost of my suit, the price of it should be *increased* by an amount to be fixed by drawing a number from a hat. If x stands for the numbers so drawn, instead of paying 20 pounds, I would have to pay $20+x$ pounds. As I had £50 to begin with, I should be left with $50-(20+x)$ pounds. The instructions corresponding to $50-(20+x)$ are 'Think of a number; add 20; subtract the result from 50.' If we experimented, we could arrive at the following table:

Think of a number	1	2	4	7	9
Add 20	21	22	24	27	29
Subtract result from 50	29	28	26	23	21

The larger the number I think of, the less the final answer is. This suggests subtraction, and, in fact, if we take the example of the bottom row, we see that each number can be obtained by subtracting the number in the top row from 30. If x represents the number thought of, the rule behind the bottom row is $30-x$. It thus seems that $50-(20+x)$ is the same as $30-x$.

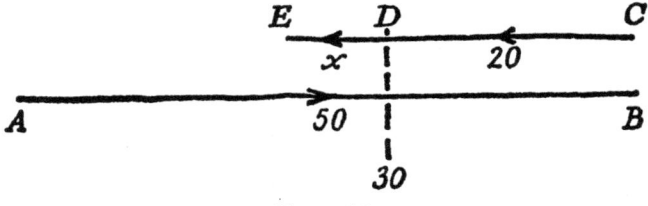

Figure 86

This result is reasonable, and can be pictured as in Figure 86. The point A shows me with nothing; B shows me with 50 pounds. The CD represents the 20 pounds I expect to pay for my suit, which will leave me at D, with 30 pounds. If my tailor persuades me to gamble, and I have to pay more than £20, I shall be left

154

A Miniature Problem in Design

with less than £30. The line *DE* shows how much more than £20 I pay, and it also shows how much my final balance will be below £30. The point *E* represents $30-x$, and it also represents $50-(20+x)$.

We now have two results:

$$\textit{Result (i)} \quad 50 - (20-x)$$
$$= 50 - 20 + x$$
$$\textit{Result (ii)} \quad 50 - (20+x)$$
$$= 50 - 20 - x$$

These results look much alike and are easily confused. Whenever in doubt, try particular numbers for x. You will soon see if you have made an error.

Result (i) is all that we need for the simple Bear Problem already considered. Result (ii) is needed if we are to see the effects of repeated subtraction (as in questions (6), (7), (8), (12), on page 150. It is also needed for Bear Problems with more than 3 posts.

Consider the following instructions. 'Think of a number; subtract it from 10; subtract the result from 30; subtract the last result from 70.' We can experiment and make a table such as that below. By examining this table, we can see a rule for each row of the table. These rules are indicated at the right. In teaching children, we would, of course, ask the pupils to guess these rules from the numbers in the table.

						Rule
Think of a number	0	1	2	3	4	x
Subtract from 10	10	9	8	7	6	$10-x$
Subtract result from 30	20	21	22	23	24	$20+x$
Subtract from 70	50	49	48	47	46	$50-x$

In using algebra, we wish to be able to pass directly from each rule to the next, without having to work out particular examples, and guess the rule for each row. Results (i) and (ii) allow us to do this. We translate the instructions into algebra as follows. 'Think of a number'; x. 'Subtract it from 10'; the result is $10-x$. 'Subtract the result from 30.' This gives $30-(10-x)$. Using the

155

pattern of result (i), this expression may be replaced by $30-10+x$, which is $20+x$. 'Subtract from 70.' We get $70-(20+x)$, and the pattern of result (ii) indicates that this is $70-20-x$, which is $50-x$.

If we were to consider instructions in which even more successive subtractions were involved, we would find that results (i) and (ii) were sufficient to guide us through all the work. The chain of instructions, however long, could always be replaced by a single subtraction or a single addition, as you probably discovered from question (12) on page 150.

BEARS PROBLEM WITH FIVE POSTS

The Bears Problem can be posed with any number of posts. In Figure 87 we have five gates attached to five posts $A, B, C, D, E,$

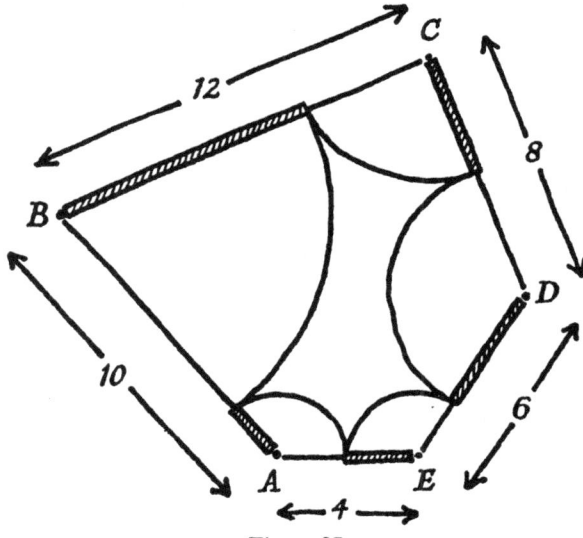

Figure 87

at distances 10, 12, 8, 6, 4. How big should the gate at A be? If it is x, the gate at B must have size $10-x$, so that side AB can be closed. In order to close BC, gate C must have size $12-(10-x)$

A Miniature Problem in Design

which (by result (i)) is $12-10+x$ or $2+x$. To close CD, gate D must have size $8-(2+x)$ which (by result (ii)) is $8-2-x$ or $6-x$. Finally to close DE, gate E must have the size $6-(6-x)$ which is $6-6+x$, or simply x. Now the gate at A and the gate at E have to fill the side AE, of length 4. Each of these gates is of size x, so $2x=4$ and $x=2$. We follow round the sides and see that the gates at A, B, C, D, E must be of sizes 2, 8, 4, 4, 2.

Exercises

How big should the gates be if the distances between the 5 posts were as specified below?
  1. 10, 11, 12, 13, 14
  2. 10, 10, 10, 10, 8
  3. 20, 30, 24, 40, 36
  4. 40, 50, 60, 55, 35
  5. 2, 3, 2, 4, 4

THE BEARS PROBLEM WITH FOUR POSTS

There is an essential difference when the number of posts is even. This difference can be discovered by experimenting with the following two problems:

Problem 1. There are 4 posts, A, B, C, D. The distances are 10, 12, 10, 16 as shown in Figure 88. If the gate at A is of size 7, how

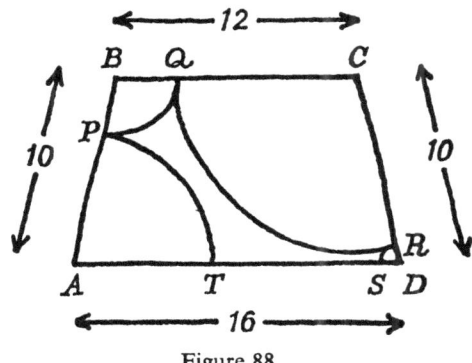

Figure 88

157

large will the gap *TS* be? What is the effect on this gap of making the gate at *A* larger or smaller? If gate *A* is of size *x*, how large will the gap be?

Problem 2. This is like problem (1), except that the distances are now as follows: *A* to *B*, 14; *B* to *C*, 12; *C* to *D*, 14; *D* to *A*, 16. Find how large the gap *TS* is for the following sizes of gate *A*: 3, 4, 5, 6, 7.

Exercises

Investigate what happens when the distances between the posts are as follows:

  (i)   10, 11, 10, 13
  (ii)   9, 10, 11, 10
  (iii) 10, 13, 10, 7
  (iv)   4, 5, 6, 7
  (v)   4, 7, 8, 5

The 5 questions above can be studied by experimenting with particular numbers, and also by algebra; in question (i) for example, if gate *A* were of size *x*, gate *B* would be of size $10-x$, gate *C* of size $11-(10-x)$ and so on. Pupils can make up many more problems of this type for themselves.

It will be seen that the exercises fall into two types, some resembling problem (1) and some problem (2). Is there any way of telling in advance to which type any exercise belongs? A pupil with a good knowledge of algebra may be able to investigate the general question, in which the distances between the posts are *a*, *b*, *c*, *d*. Similar investigations are, of course, possible with 3, 4, 5, 6, or any number of posts.

EXPLAINING A CONJURING TRICK

The following trick with cards is described in *Mathematics, Magic and Mystery* by M. Gardner, Constable, 1956. It is quite striking and does not require any sleight of hand.

The conjurer asks someone to shuffle a pack of cards. The conjurer then writes something on a piece of paper, which is then placed in some conspicuous position, where the conjurer has no chance to get at it. The conjurer puts 12 cards on a table. A spectator chooses 4 of the cards. The other 8 are put at the back of the pack. If the chosen cards are, for example, 2, 3, 5, and 7 (the suits do not matter), these numbers are added together, $2+3+5+7=17$, and the result noted. The conjurer deals out cards on top of each card. He counts from the value of that card up to 10. For example, on the 7 he would deal 3 cards, counting 'Eight, nine, ten'. When he has completed in this way 4 piles of cards on the table, he returns to the cards in his hand. The number 17 was noted earlier, in our example. The conjurer counts to the 17th card in the pack, and displays it. A spectator then examines the piece of paper, and finds the name of this card is written on it. The effect is quite astonishing to anyone unfamiliar with mathematical card tricks. It can produce a great sensation if used as part of an arithmetic or algebra lesson.

We now consider the explanation of this trick. The conjurer, after the cards have been shuffled, steals a glance at the bottom card of the pack. The name of this card is what he writes on the paper. The rest of the trick is purely automatic. Suppose, for example, this card is the Queen of Hearts. We indicate the situation in the pack by Q/51. Q indicates the Queen of Hearts; the 51 shows 51 cards above her. Now 12 cards are dealt on to the table, from the top of the pack; 4 of them remain there, the other 8 are returned to the bottom of the pack. The situation in the pack is now 8/Q/39. There are 8 cards below the Queen, and 39 (this being $51-12$) above her. So the Queen is the 40th card in the pack. We could simply count to the 40th card and produce the Queen of Hearts, in agreement with what is written on the paper. But there would then be little mystery in the trick. Our aim is to count to the 40th card, but to camouflage what we are doing, by breaking it up and making it – apparently – depend on all kinds of chance factors, such as the values of the 4 cards selected by a spectator.

It might happen that the 4 cards on the table were all tens. Addition would give $10+10+10+10=40$, the number to be noted. Since all the cards are already 10, the conjurer would not deal any cards out from the pack; the piles are already complete. He then counts to the 40th card, which, of course, is the Queen of Hearts.

Vision in Elementary Mathematics

Now we vary things a little. Suppose the 4 cards were 10, 10, 10, 9. The total is now 39, and when we come to count down the pack, at the end of the trick, we shall only count to 39. But before we do this we shall have dealt out a card on top of the 9, saying 'Ten' as we do so. We shall again arrive at the 40th card in the pack.

In the same way, if the cards were 10, 10, 10, 8 we should deal out 2 cards (with the words, 'Nine, ten'), and count 38 down the cards remaining in the pack.

If you continue to vary the conditions, you will find that any change which causes 1 more card being dealt out also causes 1 less card to be counted in the pack, so that the final result is always the same.

THE ALGEBRA BEHIND THE TRICK

It will be seen that this trick has something in common with the 'think of a number' tricks in which, whatever number you choose, the final answer is always the same. It is suggested that we might explain this conjuring trick by means of algebra, and, in fact, algebra is commonly used by anyone who wants to invent a trick of this type.

Suppose then we use a, b, c, d to represent the numbers on the 4 cards chosen by the spectator. The total of these is $a+b+c+d$; this is the number to be noted. How shall we represent the number of cards dealt on top of the card a? Consider a few examples. If the card is 7, we count 'Eight, nine, ten' and put 3 cards on it as we do so. If it were 6, we would count 'Seven, eight, nine, ten', and put 4 cards on it as we counted. The smaller the number of the card, the more cards we put on it. In fact, the rule is subtraction; on 7 we put $10-7$ cards, on 6 we put $10-6$ cards, and so on. On a we put $10-a$ cards. In the same way, we deal $10-b$ cards on top of b; we deal $10-c$ cards on top of c, and $10-d$ cards on top of d. The total number of cards we deal out may thus be written $(10-a)+(10-b)+(10-c)+(10-d)$. This is the sort of expression that gives no trouble at all to an experienced mathematician, but can be very confusing to a beginner.

A Miniature Problem in Design

Let us retreat for a moment from algebra to arithmetic, and think what this expression would mean in our first example where the cards were 2, 3, 5, and 7. We then had to deal $10-2$ cards on top of the 2, then $10-3$ cards on top of the 3, then $10-5$ cards on top of the 5, and finally $10-7$ cards on top of the 7. We are interested in the total number of cards dealt out.

We may picture these numbers as in Figure 89. In the top row we have 10 objects, 2 of which have been crossed out. The number of objects remaining (not crossed out) in the top row is $10-2$.

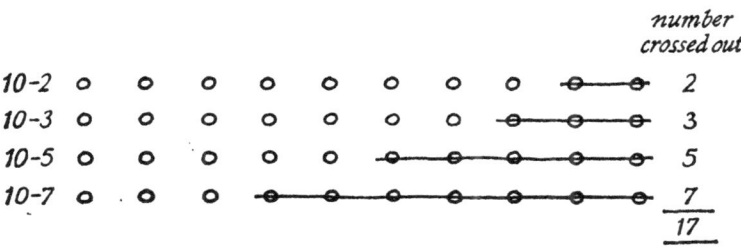

Figure 89

In the same way, $10-3$ objects remain in the second row, $10-5$ in the third row, and $10-7$ in the fourth row. The total number of objects remaining, $(10-2)+(10-3)+(10-5)+(10-7)$, represents the number of cards dealt out on to the table.

If we look at Figure 89 as a whole, we see that there were originally 40 objects, of which 2 are crossed out in the first row, then 3 in the second, then 5 in the third, and finally 7 in the fourth. The number remaining is $40-2-3-5-7$. But we also saw that the number remaining could be written $(10-2)+(10-3)+(10-5)+(10-7)$, by adding together the numbers remaining in each row. So these are simply two different ways of writing the same number. That is:

$$(10-2)+(10-3)+(10-5)+(10-7)=40-2-3-5-7.$$

There is nothing special about 2, 3, 5, and 7. The same kind of picture could be drawn for any numbers a, b, c, d and so we have:

$$(10-a)+(10-b)+(10-c)+(10-d)=40-a-b-c-d.$$

161

This result looks very reasonable. The 4 tens have been added together to make 40; a, b, c, d are still there with minus signs in front of them.

You may remember that early in the trick we noted down the number $a+b+c+d$. After the cards had been dealt on to the piles on the table, $a+b+c+d$ was the number of cards counted in the pack. So the effect is this:

Cards dealt into the piles $40-a-b-c-d$
Cards counted in the pack $a+b+c+d$

It is not hard to see that the total of these two numbers is always 40. Adding $a+b+c+d$ to $40-a-b-c-d$ puts back the a, b, c, and d that have been subtracted earlier. So we always end on the 40th card of the pack, as we wish to do.

You may notice in Figure 89 that the total number of objects crossed out is 17, which is the same as the number noted when the numbers on the cards were added together. In this example, the number of cards dealt into the piles is $40-17$, and the number counted in the pack is 17. So it is not surprising that we end at the 40th card.

ANOTHER PROBLEM OF DESIGN

In Chapter 6, in connexion with Pythagoras' Theorem, we saw various ways of fitting two squares together to make a single square. In many geometry books the diagram in Figure 90 will be found.

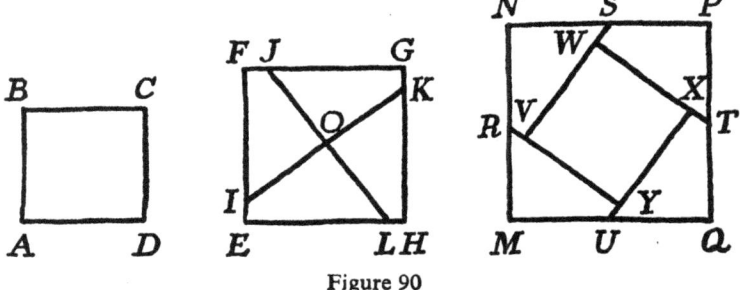

Figure 90

A Miniature Problem in Design

The third square has been made from the first two.

I once wanted to demonstrate this dissection to a class. I had remembered the general idea, as shown in the drawings above. I wanted the first square to be 6×6 and the second to be 8×8, so as to make the third square 10×10, since 36+64=100. My problem was how to place the points *I, J, K, L* on the sides of the second square. If you choose any point *I* on the side *EF*, and make *FJ*, *GK*, and *HL* the same length as *EI*, you will find that the 4 pieces of the square *EFGH* can always be fitted together to make a square *MNPQ* with a square hole, *VWXY*, in the middle. But, of course, the size of this square hole will depend on the place chosen for *I*. How should we place *I* so as to make *VWXY* a square of just the right size for *ABCD* to fit into?

This problem looks very much harder than it really is, mainly because the pieces move in such a way that we lose sight of where each has gone to.

The decision to be made is where to put the point *I*. Very well, suppose *I* placed so that the distance *EI* is *a* and *IF* is *b*. What size of hole, *VWXY*, will be the result? We place a heavy black strip along every line of length *a*, and a white strip along every line of length *b*. The pieces then move as shown in Figure 91.

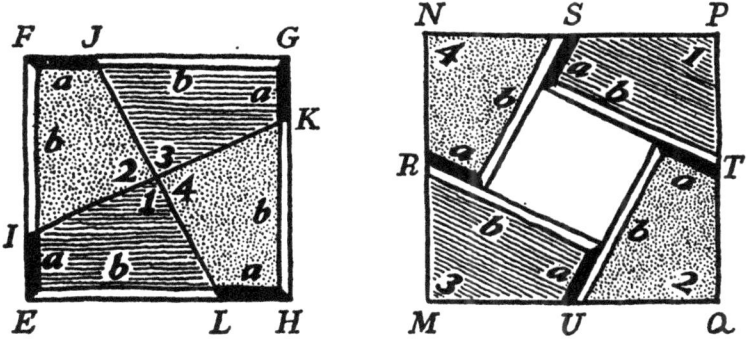

Figure 91

The distance *VW* is the side of the square hole. As the distance *VS* is *b*, and the distance *WS* is *a*, the distance *VW* is *b−a*, a very simple result indeed.

163

The side *EH* of the middle square is $a+b$, for it consists of two parts of lengths a and b.

I wanted *EH* to be 8 and *VW* to be 6. This meant I had to choose a and b so that $b-a=6$ and $a+b=8$. The equations here are of the type considered in Chapter 3, but without using that chapter it is easy to guess the answer: $b=7$, $a=1$. If you make *EI* 1 inch long, and *IF* 7 inches, and cut out paper or cardboard to this specification, you will find that everything works out as desired.

Several more exercises in algebra and geometry may be suggested by a study of this dissection. These will not be gone into now. The purpose of this section has simply been to give another illustration of how a little algebra may be used in settling the details of some design.

CHAPTER EIGHT

Investigations

ONE of the great difficulties in playing chess is to play slowly enough. Too often, you make a move and immediately afterwards you see that it was a bad move. You realize you should have seen this before making the move.

In learning algebra a similar difficulty occurs. Time and again pupils hand in work that is plainly ridiculous. If they had thought what it meant, they would have seen that it was ridiculous. A student of algebra must acquire the habit of weighing his words; he must not write a statement unless he is satisfied that it is true. Some mistakes in algebra seem to be made by pupils of all ages and all countries; these mistakes have almost the standing of international mathematical heresies. They seem to come about in some such way as the following. Teachers are concerned to convey to their pupils the principle we first met in Chapter 3, that if $x+y$ is to be (say) doubled, the result is $2x+2y$. For a long time, the pupils make mistakes about this; they write an answer such as $2x+y$. At long last they learn to do this computation correctly. The idea is impressed on their minds: if you want twice $x+y$ you must add twice x to twice y. Firm in this belief, the pupils then meet the problem of squaring $x+y$. 'What could be simpler?' they think. 'We must add the square of x to the square of y.' The same principle is applied to square roots; evidently (they reason) the square root of $x+y$ should be the square root of x added to the square root of y. Again, what is 120 divided by $x+y$? Confidently they answer that it is 120 divided by x added to 120 divided by y. Unfortunately, all three arguments are based on false analogy. All three answers are incorrect and indeed are complete nonsense

Now here a teacher of algebra finds himself in something of a dilemma. Algebra is the language of much of mathematics, science, and engineering; it is desirable that pupils should be fluent in this language – that is, they should be able to do algebra quickly. But if they are to weigh their words and continually to ponder the correctness of their statements, surely this means that they will only be able to do algebra slowly?

If we had to choose one horn of this dilemma, it is clear which one we would take. It is better to write correct statements slowly than incorrect statements fast.

But, in fact, looking at the process of education as a whole, we are not seriously perplexed by this dilemma. We have already emphasized, in Chapter 4, that the main purpose of early teaching is to enable the pupil to distinguish truth and falsehood, sense and nonsense, for himself. At this stage, the work is of necessity deliberate, exploratory, unhurried. If in his early years a pupil many times reasons out for himself that $(x+y)^2$ cannot possibly be x^2+y^2, in later years he will simply *know* this, immediately and without any need for reflection.

It is easy enough to see that $(x+y)^2$ is different from x^2+y^2. If, for example, x were 3 and y were 4, $x+y$ would be 7. How does 7^2 compare with 3^2+4^2? We can picture 3^2+4^2 as in Figure 92.

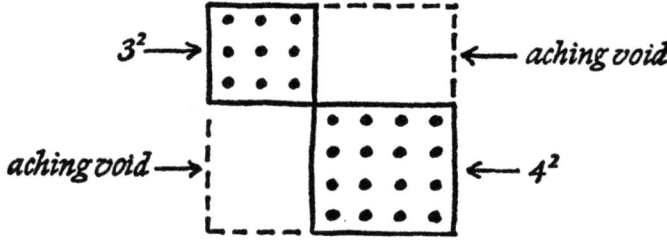

Figure 92

Putting together the square of 3 and the square of 4 falls far short of making the square of 7. In fact, the figure shows what is missing. We have to fill 2 rectangles, each of which we associate with 3×4. In fact $7^2=3^2+4^2+2\times3\times4$, corresponding to the general result $(x+y)^2=x^2+y^2+2xy$ which we have already met in Chapter 6.

In this chapter, some investigations are given, which children can carry out and enjoy. None of them use any technique more

Investigations

elaborate than the drawing of diagrams similar to Figure 92 above. Most of them help children to build up the conviction that $(x+y)^2$ is not x^2+y^2 – something for which the children's future teachers and professors should be grateful.

AN INVESTIGATION OF SQUARES

In this investigation, the pupils are first asked to calculate the squares of the numbers from 0 to 20. They are then to see what they notice.

To an adult, the interest seems concentrated in the second part – the search for a pattern in the numbers after they have been calculated. But I have known a class to cheer when the first part of the work was announced. The only explanation I can find is based on a remark of A. P. Rollett, that children enjoy making complete collections. They would far sooner (he says) make a table showing the prime factors of all the numbers up to 100 than find the prime factors of a few numbers chosen at random by the teacher. The making of tables is an excellent way of improving accuracy in computation. It is one of the defects of the conventional arithmetic texts that the exercises are totally disconnected. The answer to one question throws no light on the reasonableness of the answer to the next. No theme can be discerned. But a task involving many calculations based on a single idea often provides its own check on accuracy. For example, once when I had asked a class to make a table of squares, a girl held up her hand and asked, 'Shouldn't 14 fourteens be *more* than 13 thirteens?' She had made a slip in finding 14×14 and had reached the answer 136. Normally, she would never have noticed that anything was wrong. But in this context, she observed the steady growth of the square numbers, . . . 49, 64, 81, 100, 121, 144, 169, and then, to her surprise, found the trend reversed: 169 was followed by the smaller number 136.

When the pupils have made the table:

x	0	1	2	3	4	5	6	7	8	9	10 ...
x^2	0	1	4	9	16	25	36	49	64	81	100 ...

they may observe a variety of patterns. There is one particular pattern we want them to notice, but, of course, they do not know what it is. We have to be careful not to get cross with the children because they are not thought readers! Any true observation they make must be praised, even if it is not the particular one we hoped for. For example, they may notice that the square numbers 0, 1, 4, 9, 16 . . . are alternately even and odd. This is perfectly true, though not very surprising, since the square of an even number must be even, and the square of an odd number odd. Once a boy called out, 'The first two are the same, but the others are different.' This somewhat cryptic remark was meant to convey a perfectly correct observation; the first column contains 0 and 0, equal numbers; the second contains 1 and 1, again equal; but in all other columns two distinct numbers appear. In effect the boy had said that $x^2 = x$ for $x = 0$ and $x = 1$, and for no other value.

Incidentally, most remarks made by children consist of correct ideas very badly expressed. A good teacher will be very wary of saying 'No, that's wrong.' Rather, he will try to discover the correct idea behind the inadequate expression. This is one of the most important principles in the whole of the art of teaching.

After a little while, children are almost sure to notice the regular way in which the square numbers increase. From 0 to 1 is a step of 1, from 1 to 4, a step of 3, from 4 to 9 a step of 5. If we indicate these steps, our table comes to have the following appearance:

x	0	1	2	3	4	5	6	7	8	9	10
x^2	0	1	4	9	16	25	36	49	64	81	100
Changes		1	3	5	7	9	11	13	15	17	19

The odd numbers appear in order in the bottom row. Children will usually rest content with this observation, but, in fact, there is still much to discover. Which odd number comes at any particular place? For example, $30^2 = 900$. If I work out 31^2, I expect it to exceed 900 by *some* odd number. Can I predict which odd number it will be without taking the trouble to calculate 31^2? There are two ways in which this question might be answered.

(i) The most natural way is to examine the table again. We pick on some particular odd number in the bottom row, say 7. Where does it occur? It gives the change between the square of 3 and the square of 4. We can hardly say these numbers, 7, 3, 4, without thinking that 7 is $3+4$. We test this by taking a few more samples. What is the change between the square of 7 and the square of 8? It is 15, which is $7+8$. The idea seems to work well. The odd number in the bottom row is found by adding together 2 numbers in the top row, as here:

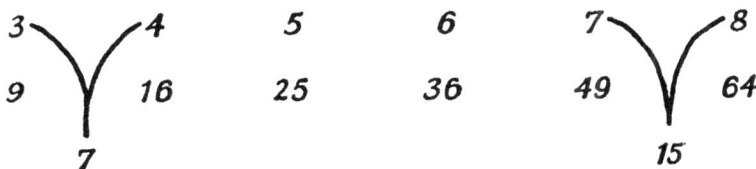

(ii) A second method might occur to someone who had practised the guessing games of Chapter 4. If we make a table showing the odd number that occurs *immediately after* any number x in the top row we obtain the following:

x	0	1	2	3	4
Change	1	3	5	7	9

The rule which lies behind this table may then be guessed – double and add 1.

The most interesting situation arises in a class where one pupil uses method (i) above and another uses method (ii). For they will then produce different answers, both of which are, in fact, correct. The teacher should be very careful not to give a verdict. The pupils should argue until they eventually agree that both methods yield a correct result. For this is one of the central themes of algebra – that two procedures, apparently different, can, in fact, lead to the same answer. One purpose of algebra is to determine when two procedures are equivalent in this way. Of two equivalent procedures, one may be much more complicated than the other: one task of algebra is, when a procedure is given, to select from all procedures that yield the same final result, the quickest and simplest. In our present investigation, there is little to choose so far as simplicity is concerned. If we want to find the change from 30^2 to 31^2, the first method tells us to find $30+31$, which is 61; the second method tells us to double 30 and add 1, again giving the answer 61. So $31^2 = 900 + 61 = 961$.

Having these methods, it is natural to express them in the language of algebra. What is the change as we go from the square of any number, n, to the square of the next whole number? First, we have to translate into algebra 'the next whole number'. To pass from 3 to 4, or from 7 to 8, or from 30 to 31, we add 1. So the next number after n is $n+1$.* Now, what is the change? The first method says, 'Add n to $n+1$'. This gives us $2n+1$. The second method says, 'Double n and add 1'. This also gives us $2n+1$. We have thus demonstrated that the two methods are, in fact, equivalent.

We have found that the change in going from n^2 to $(n+1)^2$ is $2n+1$. If we add this change to n^2, we must get $(n+1)^2$. It follows that $(n+1)^2 = n^2 + 2n + 1$.

As has been mentioned earlier, in arithmetic we never leave a question in a form as $100 + 20 + 1$. If an expression contains a $+$ sign, we work it out. Pupils may ask whether we cannot 'work out' $n^2 + 2n + 1$. No great harm is done here by a certain amount of 'telling'; we may tell the children that there seems to be no way of shortening this expression. Or we may ask the class to suggest what a shorter form might be; they then test these suggestions by trying particular numbers, and find that none of them work. It may also amuse the class if you give them the following instructions; 'Write down these numbers: seven million; two thousand; three hundred. Add them up. Read out your answers.' It will appear here that the answer simply repeats the question. The question was, 'What is seven million, two thousand, and three hundred?' The answer is, seven million, two thousand, and three hundred.' There is no shorter form.

We shall, of course, emphasize that $(n+1)^2$ is *not* $n^2 + 1$, as they might guess.

We can explain, and at the same time emphasize, this result by getting the pupils to illustrate it. Take, for example $n=3$. We picture 3^2 as a square, 3 rows of 3. According to our result, to make this into 4^2 we have to add 2 threes and 1. Where should the 2 threes and the one 1 be put, to make a square with 4 rows of 4? Children readily give the solution (Figure 93).

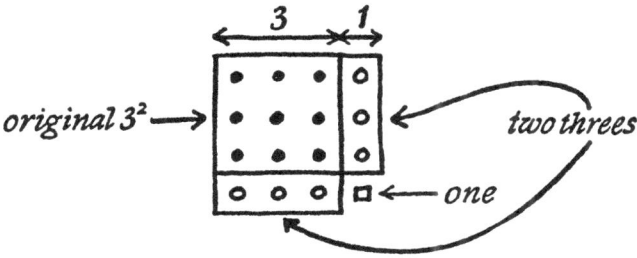

Figure 93

After doing other particular cases, they will equally readily devise the illustration (Figure 94) for the general case (compare Chapter 5).

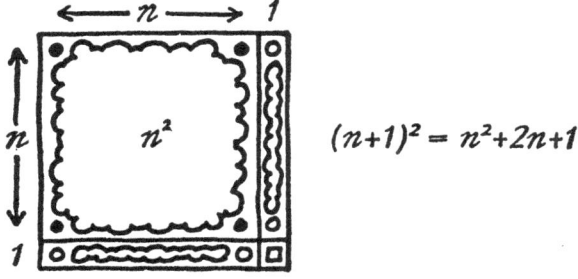

$$(n+1)^2 = n^2 + 2n + 1$$

Figure 94

Exercises

1. 101^2 is larger than 100^2 by 201, as can be found by either of the methods just discussed. Since $100^2 = 10,000$, 101^2 must be 10,201. This way we have worked out 101^2 by mental arithmetic. Use the same argument to find (i) $1,001^2$; (ii) $10,001^2$; (iii) $100,001^2$; (iv) $1,000,001^2$; (v) 201^2; (vi) $2,001^2$; (vii) $3,001^2$; (viii) 301^2; (ix) 31^2; (x) 21^2.

2. 20^2 is larger than 19^2 by 39. As $20^2 = 400$, 19^2 must be 39 less than 400, that is $19^2 = 361$. How much is 100^2 larger than 99^2? Use this result to find 99^2.

171

Vision in Elementary Mathematics

3. $12^2=144$. If you write both numbers backwards, you get the true statement $21^2=441$. Find other examples of this type of behaviour.

4. Figure 94 illustrates the meaning of $(n+1)^2$. Draw a similar picture of $(n+2)^2$, and use it to complete the following statement: $(n+2)^2=n^2+ \ldots + \ldots$. Do the same with $(n+3)^2$ and $(n+4)^2$. What do you notice about the results? What do you think $(n+10)^2$ should be? If we use $(n+a)^2$ for the square of n plus *any number*, you can express what you have noticed writing $(n+a)^2=n^2+ \ldots + \ldots$.

5. By how much is n^2 larger than $(n-1)^2$? Method (i) of the text gives the answer most easily.

6. Could the answer to question (5) be used to find an expression for $(n-1)^2$, corresponding to the expression n^2+2n+1 for $(n+1)^2$? (Result (i) of Chapter 7 resembles what we use here.)

MORE TRICKS

The following trick is less obvious than the 'Think of a number' tricks in Chapter 4. Write down any 4 numbers that come together when you count, for example 7, 8, 9, 10. Multiply the middle numbers together: $8\times9=72$. Multiply the end numbers together: $7\times10=70$. Find the difference between these two results: $72-70=2$. This difference always turns out to be 2.

One can present this trick as an exercise in arithmetic. Ask the pupils to work out the following:

$$2\times3-1\times4=$$
$$3\times4-2\times5=$$
$$4\times5-3\times6=$$
$$5\times6-4\times7=$$

They will discover that the answer is 2 in each case. We must, of course, remind them that $5\times6-4\times7$ means '5 sixes, subtract 4 sevens', as pointed out in Chapter 5. A discussion can arise from these 'coincidences'. What do they expect $24\times25-23\times27$ to be? They say 2. Work it out and see if this guess is correct. How would they fill in the spaces to make the statements below correct?

Investigations

$$9 \times 10 \ - \ 8 \times \ldots = 2$$
$$13 \times 14 \ - \ldots \times \ldots = 2$$
$$17 \times \ldots - \ldots \times \ldots = 2$$
$$98 \times \ldots - \ldots \times \ldots = 2$$
$$\ldots \times \ldots - \ 9 \times \ldots = 2$$
$$\ldots \times \ldots - \ 15 \times \ldots = 2$$
$$\ldots \times \ldots - 243 \times \ldots = 2$$

This trick may be used in many ways and at many levels. We will illustrate these in turn, with the simplest uses first.

In teaching arithmetic, this trick could be used as a way of getting children to perform a large number of multiplication exercises without becoming bored. The trick aspect provides an automatic check on accuracy; if a child reports that he has found a number for which this trick fails, he would be well advised to check his working – the trick does, in fact, work for all numbers, though the class should not be told this at the outset.

Still at the level of arithmetic, children can experiment with variations on this trick, as shown below.

Calculate the following, and see what you notice.

A1	$2 \times 4 - 1 \times 5 =$	
	$3 \times 5 - 2 \times 6 =$	
	$4 \times 6 - 3 \times 7 =$	etc.
A2	$2 \times 5 - 1 \times 6 =$	
	$3 \times 6 - 2 \times 7 =$	
	$4 \times 7 - 3 \times 8 =$	etc.
A3	$3 \times 4 - 1 \times 6 =$	
	$4 \times 5 - 2 \times 7 =$	etc.
A4	$2 \times 2 - 1 \times 3 =$	
	$3 \times 3 - 2 \times 4 =$	etc.
A5	$3 \times 3 - 1 \times 5 =$	
	$4 \times 4 - 2 \times 6 =$	etc.
B1	$2 \times 6 - 3 \times 4 =$	
	$3 \times 7 - 4 \times 5 =$	
	$4 \times 8 - 5 \times 6 =$	
	$5 \times 9 - 6 \times 7 =$	etc.

173

B2 $3 \times 8 - 4 \times 6 =$
           $4 \times 9 - 5 \times 7 =$
           $5 \times 10 - 6 \times 8 =$ etc.

B3 $4 \times 9 - 6 \times 6 =$
           $5 \times 10 - 7 \times 7 =$
           $6 \times 11 - 8 \times 8 =$ etc.

B4 $3 \times 4 - 1 \times 5 =$
           $4 \times 5 - 2 \times 6 =$
           $5 \times 6 - 3 \times 7 =$ etc.

c1 $2 \times 4 - 1 \times 3 =$
           $3 \times 5 - 2 \times 4 =$
           $4 \times 6 - 3 \times 5 =$
           $5 \times 7 - 4 \times 6 =$ etc.

c2 $2 \times 5 - 1 \times 4 =$
           $3 \times 6 - 2 \times 5 =$
           $4 \times 7 - 3 \times 6 =$ etc.

c3 $2 \times 6 - 1 \times 5 =$
           $3 \times 7 - 2 \times 6 =$
           $4 \times 8 - 3 \times 7 =$ etc.

D1 $3 \times 3 - 1 \times 2 =$
           $4 \times 4 - 2 \times 3 =$
           $5 \times 5 - 3 \times 4 =$
           $6 \times 6 - 4 \times 5 =$ etc.

D2 $2 \times 5 - 1 \times 3 =$
           $3 \times 6 - 2 \times 4 =$
           $4 \times 7 - 3 \times 5 =$
           $5 \times 8 - 4 \times 6 =$ etc.

Each of these questions ends with 'etc'. In each question, enough is given to establish the pattern. In each row, it will be noticed, each number is 1 more than the corresponding number in the row above. If, in the last question, D2, we wished to continue the pattern, we would take the last row printed, $5 \times 8 - 4 \times 6$, and increase each number in it by one. This would give us $6 \times 9 - 5 \times 7$.

Investigations

Another way of putting the same thing is to point out the behaviour of the numbers in each column. The first column of D2 contains 2, 3, 4, 5. If we wished to continue this column downwards, we would naturally write 6, 7, 8, 9 . . . . In the same way, the other columns can be continued downwards.

Children easily see the pattern of the columns. They should have their attention drawn to the pattern of the rows. For example, in D2, we find the row $5 \times 8 - 4 \times 6$. The smallest number occurs in the third place, namely 4. The first number 5, is 1 more than 4. The last number 6, is 2 more than 4. The second number is found by adding 4 to the smallest number. So in D2 we should meet the row:

$$101 \times 104 - 100 \times 102$$

if we went down far enough.

Children should be encouraged to experiment, by writing down any 4 numbers and building on them questions similar to those above.

They will notice that the questions fall into types. These types are indicated above by letters A, B, C, D. In questions A1 to A5, the effect is as in the trick originally described. Questions B1 to B4 would not serve for a trick of the same type, but something else happens which can be observed. The answers to C1, C2, and C3 exhibit a common feature, slightly different from that of type B.

It is natural to ask the question: if we select any 4 numbers, is there any way of making sure *in advance* that they will give a trick of type A? For example:

$$11 \times 14 - 1 \times 24$$
$$12 \times 15 - 2 \times 25 \qquad \text{etc.}$$

does in fact belong to type A, and gives the answer 130 throughout. How did I know how to choose these numbers? The answer, once found, is simple enough, but it is not too easy to guess. It will appear automatically when, in the next section, we explain our tricks by means of algebra.

EXPLANATION BY PICTURES

Naturally, with any trick, children will ask, 'Why does it work?' We consider our first trick which used $8 \times 9 - 7 \times 10 = 2$. This

175

dealt with 4 numbers, such as 7, 8, 9, 10, that come together when we count. We can begin with 'any number we like', n for short. In our example n is 7. Next we want to translate into algebra 'the number that follows n when we count'. We might invent a special sign for this, such as Fn where F is short for 'what follows'. This, however, would be extremely inconvenient. In elementary algebra we try to work with the usual signs of arithmetic; the only new things we bring in are letters such as a, n, x to stand for numbers chosen by different people, the heights in feet of men and their sons, and so forth. Children do not always find it easy to translate into algebra 'the number that follows n', so we must discuss this as a serious problem. Suppose you and I play a guessing game; you are to say any whole number; I have decided to answer the number *that follows yours*. The record of our game will look something like this:

You say:	5	9	15	2
I answer:	6	10	16	3

By this time you would have guessed my rule; you would probably say that I always added 1 to your number. This translates into algebra without difficulty; you say n; I answer $n+1$.

In the same way, the number that follows $n+1$ is $n+2$; after $n+2$ comes $n+3$. So if, instead of beginning with 7, we begin with any number n, the 4 numbers that replace 7, 8, 9, 10 will be n, $n+1$, $n+2$, $n+3$.

The trick next told us to multiply together the 2 middle numbers. Instead of 8×9 we shall now have $(n+1)(n+2)$ which, as we saw on page 170 can be pictured as in Figure 95. The trick also tells us to multiply the end numbers. Instead of 7×10 we shall now have $n(n+3)$, which is also pictured in the same figure.

We noticed that 8×9 was 2 more than 7×10, and the trick says that this happens whatever number you begin with. That is, $(n+1)(n+2)$ should always be 2 more than $n(n+3)$. In the picture, this means that the rectangle on the left should contain just 2 more objects than the rectangle on the right. But it is pretty clear that this is so. The rectangle on the right contains a square and 3 long thin pieces. The rectangle on the left also

176

contains a square and 3 long thin pieces (one of these is lying on its side); but, in addition, it contains 2 objects at the bottom

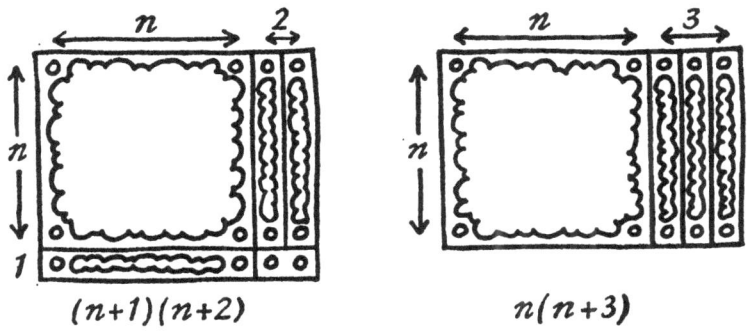

$(n+1)(n+2)$ $n(n+3)$

Figure 95

right corner. In fact, since the square contains n rows with n objects in each, it represents n^2. The long thin pieces represent n. Using the pictures we may write:

$$(n+1)(n+2)=n^2+3n+2$$
$$n(n+3)=n^2+3n$$

Clearly the first expression, n^2+3n+2, is 2 more than the second expression n^2+3n.

Exercise

Draw pictures to illustrate A1, A2, A3, and A4 of page 173. Adapt the argument in the section above to explain each of these tricks in terms of algebra.

Examine also the questions of type B, C, and D. Draw pictures for these, and state in terms of algebra what they show. (Answers at back of book. A particular question, B4, is studied in detail in the next section.)

A PATTERN BEHIND THE TRICKS

If we carry out the arithmetic in question B4 we obtain the following answers:

$$3\times4-1\times5=7$$
$$4\times5-2\times6=8$$
$$5\times6-3\times7=9$$
$$6\times7-4\times8=10$$
$$7\times8-5\times9=11$$

The answers are not all the same number, as happened with type A, but they do behave in a regular way. We see the answers 7, 8, 9, 10, 11 in counting order.

In the last row written above, we worked out $7\times8-5\times9$. The numbers appearing here are 5, 7, 8, 9. We could get these numbers by counting on from 5, which would give us 5, 6, 7, 8, 9 and then crossing out 6. If instead of 5 we began with 100, counting would give us 100, 101, 102, 103, 104 and crossing out the second number, 101, would leave us with 100, 102, 103, 104. If we began with any number, n, counting would give us n, $n+1$, $n+2$, $n+3$, $n+4$, and removing the second number would leave us with n, $n+2$, $n+3$, $n+4$. If we look back at our particular case 5, 7, 8, 9 we see that we multiply the middle numbers 7×8, and subtract the result of multiplying the end numbers 5×9. If we apply the same procedure to n, $n+2$, $n+3$, $n+4$ we get $(n+2)(n+3)-n(n+4)$ as the algebraic expression that lies behind question B4. In fact, if you replace n in turn by 1, 2, 3, 4, 5 you can check that this expression does give us exactly the arithmetical expressions standing at the beginning of this section.

What we have just done is the hardest part of this question. Children may need to be helped by questions and a little advice. Where does the smallest number occur in $7\times8-5\times9$? It is the third number, 5. It is usually simplest to work from the smallest number. We advise the children to take the number in the third column as n, the number you think of. How do the numbers in the first column compare with those in the third column? They are 2 more. What is the rule for them? It is $n+2$. In the same way we find the rule $n+3$ for the second column and $n+4$ for the fourth. We write these at the head of the various columns.

$$
\begin{array}{cccc}
n+2 & n+3 & n & n+6 \\
3 \times & 4 & -1\times & 5 \\
4 \times & 5 & -2\times & 6
\end{array}
$$

Now we have only to bring into the algebra the operations of multiplication and subtraction that occur in the arithmetic. We probably have to remind our pupils of the need to introduce brackets, as discussed in Chapter 5. We also remind them that the multiplication sign, $\times$, is not very convenient for algebra, and is usually replaced by a dot. We have now reached an acceptable algebraic expression $(n+2).(n+3)-n.(n+4)$. Some pupil will probably ask, 'Can't you leave the multiplication dots out altogether?' and we agree; you can. It is purely a matter of taste whether you put them in or not. You will be understood (by mathematicians) either way.

The rest of the work is straightforward. We must draw pictures of $(n+2)(n+3)$ and $n(n+4)$. No new idea is needed here. We use the ideas of Chapter 5 and obtain the following diagram (Figure 96).

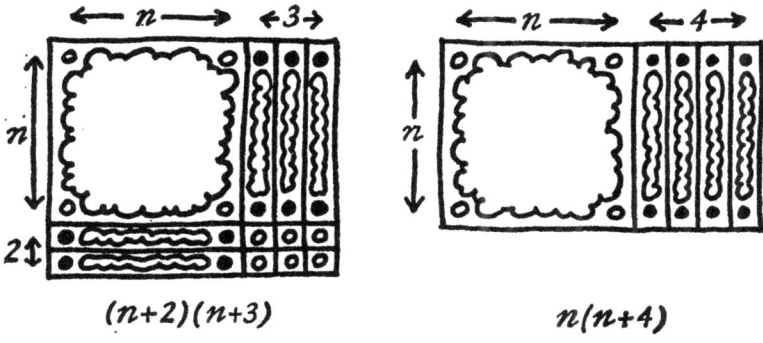

(n+2)(n+3)   *n(n+4)*

Figure 96

On the left, we have enough material to make the rectangle on the right and have a long thin piece and 6 objects left over. The thin piece represents n, so we may say that the quantity shown on the left exceeds that on the right by $n+6$. We may write this in the form of an equation:

$$(n+2)(n+3)-n(n+4)=n+6$$

This result agrees with our original arithmetic. It says the answer will be 6 more than the number originally thought of. And

Vision in Elementary Mathematics

this we observe; if, for example, we think of 5 for n, we get $7 \times 8 - 5 \times 9 = 11$ and 11 is indeed 6 more than 5.

We could arrive at the result above in a slightly more formal way. We can use Figure 95 above to multiply out $(n+2)(n+3)$ and $n(n+4)$. From the picture we see $(n+2)(n+3)=n^2+5n+6$ and $n(n+4)=n^2+4n$. The first of these exceeds the second by $n+6$.

While it is always possible to multiply out an expression such as $(n+2)(n+3)$ by drawing a picture, this is not the quickest way of obtaining the answer. The picture should be used whenever a pupil is in difficulty, and whenever the teacher is in doubt whether the pupil really understands what is happening. But once understanding has been reached, a quicker procedure may be followed. If we examine the result above $(n+2)(n+3)=n^2+5n+6$, we see that the answer, n^2+5n+6 contains the numbers 5 and 6, both of which are related in a simple way to the numbers 2, 3. We can ask the pupils, 'How would you get 5 from 2 and 3? How could you get 6 from 2 and 3?' We must not build too much on a single example; it might be merely a coincidence. The following exercise provides further evidence.

Exercise

By means of pictures, multiply out the following:

1.	$(n+2)(n+4)$	**2.**	$(n+2)(n+5)$
3.	$(n+3)(n+4)$	**4.**	$(n+3)(n+5)$
5.	$(n+1)(n+3)$	**6.**	$(n+1)(n+4)$

Is any simple rule suggested by the answers? (Solution is at back of book, and is also discussed immediately below.)

The results of this exercise confirm our guess from the single example. When we multiplied out $(n+2)(n+3)$ we obtained n^2+5n+6 in which 5 is $2+3$ and 6 is 2×3. In each of the other examples, we find an addition and a multiplication. This leads us to believe that, for example, $(n+10)(n+13)$ should be $n^2+23n+130$, since $23=10+13$ and $130=10 \times 13$.

If we look back at Figure 96, we can see why $2+3$ and 2×3 should occur when $(n+2)(n+3)$ is multiplied out. In the bottom right-hand corner of the rectangle that illustrates $(n+2)$ $(n+3)$

we see 2 rows with 3 objects in each – our standard picture for
2×3. If we ask how many long thin pieces occur, we see that
there are 3 of them standing up, and 2 lying down, $3+2$ in all.
There is nothing special about 2 and 3 in this connexion. It
would be very troublesome to draw in detail the picture for
$(n+10)(n+13)$ but if you imagine or sketch roughly how this
picture would appear, you will see that 10 rows of 13 objects
come at the bottom right-hand corner, illustrating 10×13; also
there are 13 thin pieces standing upright and 10 lying down,
representing $23n$ in all. (I am assuming that you have $n+10$ rows
with $n+13$ objects in each. My description would have to be
modified if you used the numbers the other way round, $n+13$
rows with $n+10$ objects in each. The final result would be the
same.)

We can now multiply out in our heads, and this enables us to
see straight away whether any particular rule gives a trick of type
A or not.

Take A2 for example. The rule behind this is $(n+1)(n+4)-
n(n+5)$. We see that $(n+1)(n+4)$ is n^2+5n+4 by adding and
multiplying 1 and 4. The second product, $n(n+5)$ is n^2+5n. The
difference between n^2+5n+4 and n^2+5n is 4. It does not matter
what number you think of, the answer is sure to be 4. This is
characteristic of type A; n does not appear in the answer. This
means that the number of 'long thin pieces' must be the same in
the 2 rectangles. In our present example, there are 5 such pieces
in each. We see $5n$ in both n^2+5n+4 and in n^2+5n. How did
these come to be there? In n^2+5n, the 5 reflects the presence of
the 5 in the original expression $n(n+5)$. In n^2+5n+4, the 5 is the
sum of 1 and 4, the numbers that appear in $(n+1)(n+4)$.

This suggests a quick way of making up examples of type A.
Suppose an example is to be $(n+3)(n+6)-n(n+ \ldots)$. What
number shall we write in the space? We write 9 because
$9=3+6$. If you work out $(n+3)(n+6)-n(n+9)$, you will find
$(n+3)(n+6)=n^2+9n+18$ and $n(n+9)=n^2+9n$. The difference
is 18, so we do have a trick of type A.

For contrast, consider $(n+1)(n+4)-n(n+2)$. Here $(n+1)
\times(n+4)=n^2+5n+4$ and $n(n+2)=n^2+2n$. The difference is
$3n+4$. This does depend on n, the number you choose. In fact,

Vision in Elementary Mathematics

$(n+1)(n+4)-n(n+2)$ is the rule behind D2:

$$2\times5-1\times3=7$$
$$3\times6-2\times4=10$$
$$4\times7-3\times5=13$$

You will notice that the answers 7, 10, 13 . . . rise by 3 each time, corresponding to the 3 in $3n+4$.

If we examine the numbers in the rule $(n+1)(n+4)-n(n+2)$, we see 1 and 4 in $(n+1)$ $(n+4)$. The sum $1+4$ is 5, which is not the same as the number 2 that occurs in the second product $n(n+2)$. So we do not expect this rule to belong to type A.

Someone who knows arithmetic but not algebra may be able to guess the procedure for spotting type A, but will not be able to explain it.

Our original trick used $2\times3-1\times4$ and you will notice $2+3=5$, $1+4=5$; the sums are the same. In A1, the top line is $2\times4-1\times5$; here $2+4=6$, $1+5=6$. In A2, we began with $2\times5-1\times6$. Here $2+5=7=1+6$. Previously I gave the example $11\times14-1\times24$ and predicted that it would lead to a trick of type A. The test was simple, $11+14=25$ and $1+24=25$; the sums were the same.

On the other hand, B1 begins with $2\times6-3\times4$. As $2+6=8$ and $3+4=7$, the sums are different and we do not expect a trick of type A.

This test is quite simple, once you know what it is. It might be hard to discover by pure observation of the arithmetical results, for the person seeking to guess this test has no clue as to the kind of thing he is looking for.

FACTORING

We observed earlier that $(n+2)(n+3)=n^2+5n+6$, where $5=2+3$ and $6=2\times3$. This is a particular example of a general method.

Sometimes we want to proceed in the opposite direction. I ask the question, 'I carried out a multiplication and arrived at n^2+5n+6. What do you think I was multiplying?'

A perfectly good answer, of course, would be 1 times n^2+5n+6.

Investigations

That might have been what I was doing. But is there any other, more interesting possibility? Can we fill the spaces in $(n + \ldots)(n + \ldots)$ in such a way as to get an answer? This way round, we are looking for two numbers which give 5 when added and 6 when multiplied. Naturally 2 and 3 is the solution.

Simple questions of this kind can be solved by pure guesswork.

Exercise

What multiplications of the type $(n + \ldots)(n + \ldots)$ would give the following results?

1. $n^2 + 7n + 12$.
2. $n^2 + 8n + 12$.
3. $n^2 + 13n + 12$.
4. $n^2 + 6n + 8$.
5. $n^2 + 8n + 15$.
6. $n^2 + 4n + 3$.
7. $n^2 + 6n + 5$.
8. $n^2 + 8n + 7$.
9. $n^2 + 25n + 24$.
10. $n^2 + 14n + 24$.
11. $n^2 + 11n + 24$.
12. $n^2 + 10n + 24$.
13. $n^2 + 2n + 1$.
14. $n^2 + 4n + 4$.
15. $n^2 + 6n + 9$.

In all the questions above, we are given an expression and asked to find its factors. It is not obvious why finding factors should be of importance. In fact, in the course of work in algebra, it often is helpful to know the factors of expressions we meet. Children as a rule are willing to do a few problems, of the type just given, simply as an exercise in guessing. A wise teacher will treat the matter with a light hand. He will not set so many exercises in factoring that the whole business becomes wearisome, but from time to time he will make sure that his pupils remember how to find factors in algebra.

An occasional complicated problem in factoring can be presented as a puzzle. How would you find the factors of $n^2 + 22n$

$+96$? Here we have to find two numbers that give 22 when added and 96 when multiplied. Guessing at random we might find this hard. There are many different numbers that multiply to give 96, for 96 is 2×48 or 3×32 or 8×12 or 4×24 – none of which give the right sum. To be sure of finding the right answer we must make a table showing the possibilities in some systematic way. If we take the numbers 1, 2, 3, 4, 5, 6, 7 . . . in turn and see which of them divide into 96, we get a table like this:

$$96 = 1 \times 96 \qquad 1 + 96 = 97$$
$$96 = 2 \times 48 \qquad 2 + 48 = 50$$
$$96 = 3 \times 32 \qquad 3 + 32 = 35$$
$$96 = 4 \times 24 \qquad 4 + 24 = 28$$
$$96 = 6 \times 16 \qquad 6 + 16 = 22$$
$$96 = 8 \times 12 \qquad 8 + 12 = 20$$

We notice 22 in the sums, so 6 and 16 are the numbers we want. $(n+6)(n+16)$ are the factors.

You may notice that the sums 97, 50, 35 . . . get steadily smaller as we go down the list. We could use this fact to save ourselves work. We could skip some of the numbers. For example, if we went straight from 1×96 to 4×24, we would see $4 + 24 = 28$, and 28 is still too large (we want 22); so we must go further on. If we skipped direct to 8×12, we would find $8 + 12 = 20$. As 20 is smaller than 22, we have gone too far; we must go back in the list, look for some earlier entry.

Instead of considering numbers that multiply to give 96, we could have begun by considering numbers that add up to 22. Our list would then begin like this:

$$22 = 0 + 22 \qquad 0 \times 22 = 0$$
$$22 = 1 + 21 \qquad 1 \times 21 = 21$$
$$22 = 2 + 20 \qquad 2 \times 20 = 40$$

It would go on until we reached:

$$22 = 10 + 12 \qquad 10 \times 12 = 120$$
$$22 = 11 + 11 \qquad 11 \times 11 = 121$$

It would stop at $11 + 11$ because the next entry would be $12 + 10$, which we have already considered as $10 + 12$.

Investigations

There would be rather a lot of work involved in making out the complete list. But here we notice that the numbers in the last column are getting steadily bigger. We want 96 in this column. 40 is too small, 120 is too large. We want somewhere in between. Suppose we try $5+17$. Then $5 \times 17 = 85$, still too small; we must go further down. We might jump to $7+15$. Then $7 \times 15 = 105$, too large; we have gone too far. There is only one entry between these; it is $6+16$ and $6 \times 16 = 96$, the desired answer.

Exercises (for those who like this kind of thing)

What would you multiply together to get the following:
1. $n^2 + 27n + 72$.
2. $n^2 + 61n + 900$.
3. $n^2 + 87n + 900$.
4. $n^2 + 58n + 840$.

CHAPTER NINE

The Routines of Algebra – I

IN Chapter 7, we saw that it was often convenient, when a problem involved many details or long chains of cause and effect, not to begin with definite numbers or sizes, but to use letters. These letters stand for numbers, but we shall only decide what numbers at a late stage in our calculations, when we have seen how all the particular decisions are going to affect the final outcome. Algebra makes it possible for us to do this, to calculate before choosing, instead of making blind decisions first, in order to have some actual numbers to work with.

It is clear that, in the course of designing some complicated object, we shall many times need to add, subtract, multiply, and divide. It is therefore necessary, if we are to carry through our planning, that we should know how to perform these basic operations of arithmetic, when some of the numbers involved are still uncertain and are represented by letters.

It should be realized that sometimes algebra seems to tell us nothing. For example, if you ask me 'What do you get when you add b to g?' I can only answer, '$b+g$', which of course, means 'b added to g' and tells you nothing new. This is like an example we considered earlier, where there were b boys and g girls in a room. What do you do to find how many children there are altogether in the room? You add the number of boys to the number of girls, or, in shorthand, you find $b+g$. There is no clever way of simplifying the work. It is already so simple that we just have to take it as it is.

But there are occasions when algebra gives us a simpler answer. Consider, purely as an illustration, the following not very practical problem. We have 240 tiles, each a foot square. We want to use all these tiles to make a path in the form of a square, 1 yard wide, as shown in Figure 97. How shall we do this?

We might begin by guessing. Perhaps the inner square should be 10 feet wide. Then the outer square will be 16 feet wide. The path is what remains if we take the inner square away from the outer one. So we shall have its area if we subtract the area of the inner square from the area of the outer square. That is, we must find 16^2-10^2, which is $256-100$ or 156. The path would require 156 tiles, and that is too small, since we have 240 tiles.

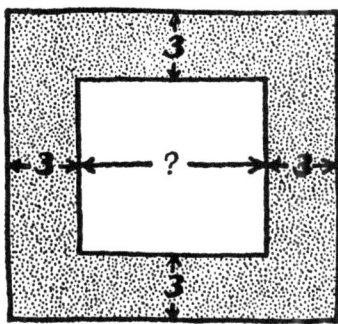

Figure 97

We might guess again, and suppose the inner square to be 20 feet across. This would lead us to 26^2-20^2 or $676-400=276$. This is too big; we have not enough tiles.

In time, by trial and error, we would hit on the answer. Trial and error is a method not to be despised. Often it will yield the solution of a problem, and it is certainly far better than sitting down and saying, 'I have not been told the rule for doing this'.

However, trial and error is a rather primitive method, and in this case it may prevent your seeing a very simple feature of the problem.

To attack the problem by algebra, we begin, of course, by supposing the inner square to be x feet wide. Since there are 3 feet of path on either side, the outer square will be $x+6$ feet wide. (Note that in each of the particular cases taken above the outer square was 6 feet wider than the inner.) The area of the outer square is accordingly $(x+6)^2$ or $(x+6).(x+6)$. We multiply this out by the methods already discussed in Chapter 8. If it helps us, we draw a picture, as explained on page 106. We find $(x+6).(x+6)=x^2+12x+36$. This is the area, in square feet, of the outer square.

187

From it we have to take away x^2, the area of the inner square. Taking x^2 away from $x^2+12x+36$, there remains $12x+36$. This is a rather simple rule. If we 'think of a number' for the width of the inner square, to get the number of tiles in the path we should multiply that number by 12, and add 36. But we want there to be 240 tiles. So we work back from 240. If, with the rule just given, we end with 240, we must have had 204 before we added 36. So we must have had 17 before we multiplied by 12.

The inner square should be 17 feet wide.

Here, the algebra showed us that there was a fairly simple way of going direct to the answer.

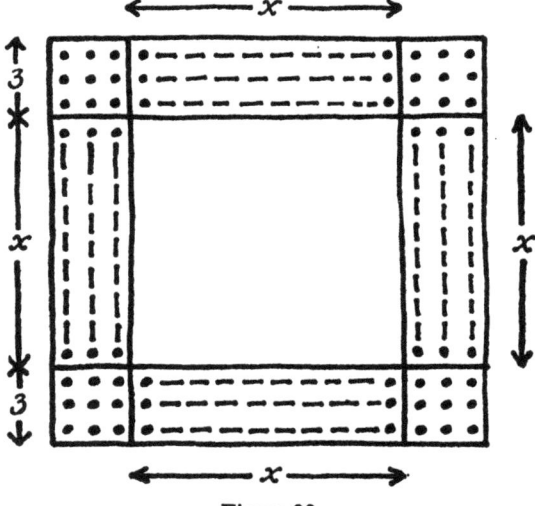

Figure 98

There is nothing mysterious about the rule, $12x+36$, for the number of tiles. In Figure 98, each dot represents a tile. You can see the 12 lines with x tiles in each, and you can see 9 tiles in each corner to make the 36. We might have noticed this way of arriving at the answer even if we had not known anything about multiplying $x+6$ by $x+6$. I find the value of routine algebra is that it tells me to look for some such simple solution in problems where I do not have the sense to see it unaided.

The Routines of Algebra I

In the previous chapters and in this discussion I have tried to give a small, and perhaps not very convincing, sample of what algebra enables us to do, and why it is worth while to learn the comparatively unexciting routine procedures for addition, subtraction, multiplication, and division in algebra. We now turn to these processes.

ADDITION

We have already noticed that arithmetic tends to give children the impression that an 'answer' should not involve any plus signs. This leads to a common type of mistake. If you ask children to add $2a$ and $3b$, they are not happy with the correct answer $2a+3b$. This answer admittedly is an anticlimax! Pupils tend to make such guesses as $5a$ or $5b$ or, on occasion, $5c$. However, if our friends, Alf, Betty, and Charles are invited each to think of a number, there just is no way of saying more simply 'Twice Alf's number added to 3 times Betty's number'. It will not (as a rule) be 5 times Alf's number (as the answer $5a$ alleges), nor 5 times Betty's (as $5b$ would imply) and there is certainly no reason to suppose it will be 5 times Charles' number, unless telepathy has been taking place. One can illustrate the meaning of $2a+3b$ as we did in Chapter 3, with boys standing on men's heads. The man can be supposed to be a feet high, and the boy b feet high. It is then clear that the picture for $3a+2b$ (the height in feet of 3 men and 2 boys) is not the same as that for $5a$ (the height in feet of 5 men), nor is it the same as $5b$ (the height in feet of 5 boys).

The difficulty here is that a question such as 'Add a to b' is rather like the riddle, 'What is the difference between an elephant and a postbox?' People rack their brains for some witty epigram contrasting the two things. Whereas in fact no such epigram is known. The riddle merely gives you a chance to insult the person when he admits that he does not know the difference between an elephant and a postbox. In the same way, when we ask children to add a and b, they start looking for some profound mathematical idea about the sum of two numbers. No such idea exists. The question is at first purely one of notation. It might prove less misleading if we asked, 'If a stands for a number Alf is thinking of, and b for a number Betty is thinking of, *how should we write in shorthand* Alf's number added to Betty's?' The pupils would then probably answer $a+b$. We might then ask, 'Is there any shorter way of writing it than $a+b$?' and some discussion might be necessary before the pupils reached the conclusion, 'No'.

We begin to reach something that looks more like a serious calculation, when we ask for the sum of $3x$ and $4x$. We have already had several ways of picturing this. We can consider x feet to be the height of a man (Chapter 3), or x to be the number of stones in a bag (Chapter 4), or the number of objects in a line, with a long thin cloud preventing us from seeing their number (Chapter 5). By any of these devices we can satisfy ourselves that $3x+4x=7x$.

By using men and their sons (Chapter 3) we could deal with additions such as:

$$
\begin{array}{ll}
\text{Add} & 2m+3s \\
& 4m+5s \\
\hline
& 6m+8s
\end{array}
$$

We could also illustrate this result by objects and clouds, as in Figure 99.

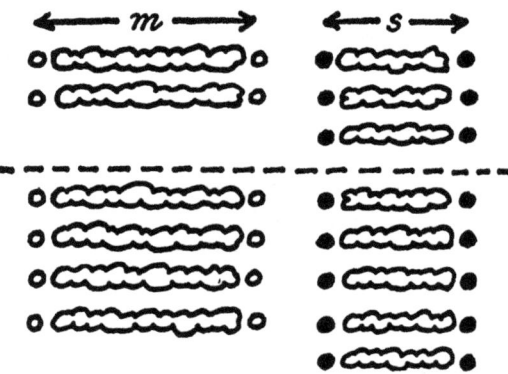

Figure 99

Above the dotted line we have $2m+3s$ objects; below the dotted line we see $4m+5s$ objects; in all we have $6m+8s$ objects.

190

The Routines of Algebra I

The desire to have an answer free from plus signs can appear very strongly when children are asked to deal with an expression such as:

$$3x^2 + 2x + 5x^3 + 6x^2 + 7x + 8$$

One is liable to get answers such as $31x^9$. The pupils have been asked to add and truly they have added every number in sight! The answer, of course, is complete nonsense. We may need to recall the notation of Chapter 5; x^2 is short for x times x and x^3 for x times x times x, where x stands for the number someone has thought of. Let us test the work above by thinking of the number 10. If $x = 10$, then $x^2 = 10 \times 10 = 100$ and $x^3 = 10 \times 10 \times 10 = 1,000$. The expression given above would thus correspond to $300 + 20 + 5,000 + 600 + 70 + 8$. There is no point in actually calculating this sum. We can see that it is not far from 6,000. Now look at the child's answer, $31x^9$. If we replace x by 10, this becomes $31 \times 10^9 = 31 \times 10 \times 10 \times 10 \times 10 \times 10 \times 10 \times 10 \times 10 \times 10 = 31,000,000,000$. Thirty-one thousand million is fantastically too big an answer for the original question. Once I asked a boy why he did it this way, and he answered, 'Father told me you should add the indices.' Now there are situations in which one should add the indices, but this is not one of them. Other children made the same mistake without any help from fathers; they just saw a lot of numbers, so they added them all up.

I found it very hard to break children of this bad habit. I found it necessary to break all such questions up into 2 parts: (i) the drawing of a picture showing what the question means; (ii) the writing of a description of your picture in the shortest possible form.

On page 114 we had pictures of x^2 and x^3.

We start with a simple example. What is a shorter way to write $x + x^2 + x$?

Part (i) Draw the picture (Figure 100).

Part (ii) What do we see in the picture? We see 1 square and 2 long thin lines. The square is a picture of x^2; each line is a picture of x. The whole picture shows $x^2 + 2x$, and this is the answer.

In the same way we might deal with $3 + 4x + 2x^3 + x^2 + 5x + x^3 + 2 + x^2$. We picture this as in Figure 101.

191

What do we see here? 3 cubes, representing $3x^3$; 2 squares, representing $2x^2$; 9 lines, representing $9x$; and 5 dots, representing 5. Thus the total may be written $3x^3+2x^2+9x+5$.

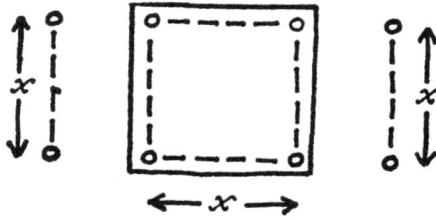

Figure 100

It is easy to make up as many exercises of this kind as you wish. The purpose of the exercises is to get the pupil out of the habit of making wild guesses, or copying what a friend has written, or using a rule that he does not understand and has remembered incorrectly, and into the habit of seeing the meaning of what he is doing. It may be helpful, while this habit is being built up, to separate the two parts of the work completely. That is to say, we do not ask the pupil to simplify $x+x^2+x$ and get the answer x^2+2x. We ask instead that he draw a picture to illustrate the meaning of $x+x^2+x$, and we go no further on that day. On another day we show him two lines and a square (as in Figure

Figure 101

100) and get him to interpret it as x^2+2x. Only when these two separate parts have been well established as habits do we put them together and go from algebra to picture to algebra. Otherwise there is a danger that pupils will short-circuit the process, and go directly from the question in algebra to the wrong answer in algebra, by some mistaken rule or rash guess.

The Routines of Algebra I

Figure 102 is an example of a little examination, to see whether pupils have learned to associate pictures with expressions in algebra. Until a pupil can score 100 per cent in such a test, it is unwise to try to teach him formal algebra, for he has no understanding of what you are talking about.

By picturing $5x^3 + 4x^2$ as the number of objects that fill '5 cubes and 3 squares', the pupil comes to see that there is no shorter way of expressing this. On the other hand, $5x^3 + 4x^3$ can be said more simply, for it is pictured as 5 cubes and 4 cubes, that is, 9 cubes; so it may be written $9x^3$. In the same way, $5x^2 + 4x^2$ corresponds to 5 squares and 4 squares, that is, 9 squares, and may be written $9x^2$.

When the pupil has begun to appreciate this, he may be encouraged to work addition exercises of the following type:

$$\begin{array}{r} 2x^3 + 5x^2 \\ 4x^3 + 6x^2 \\ \hline 6x^3 + 11x^2 \end{array}$$

It is easy to draw pictures of the two expressions that are being added. When we combine them the 2 cubes and the 4 cubes combine to make 6 cubes, the 5 squares and the 6 squares combine to make 11 squares.

In this way the pupil comes to realize that he must make separate counts of cubes and squares.

It is not easy to devise pictures for x^4 and x^5 and indeed it is not necessary to do so. If a child is able to carry out an addition such as:

$$\begin{array}{r} 8x^3 + 3x^2 + 6x + 2 \\ 4x^3 + 7x^2 + 5x + 1 \\ \hline 12x^3 + 10x^2 + 11x + 3 \end{array}$$

and he can draw a picture to illustrate it, and understands why we cannot lump the 12 and 10 and 11 and 3 together and say the answer is 36 something-or-other, then this child will probably carry out correctly an addition such as:

Vision in Elementary Mathematics

The pictures are supposed to help students to understand the meaning of the following:

$4+3$	$4 \cdot 3$	$2 \cdot 4 + 3$	$2 \cdot (4+3)$	$a+b$
$a \cdot b$	$a \cdot (b+c)$	$a \cdot b + c$	$3 \cdot x$	xy
$x+y$	x^2	x^3	$x^2 + 2x + 3$	$x^2 + x + 1$

Write against each picture the one that goes with it. The answers are all in the list just given, but not in that order. The first question has been answered already.

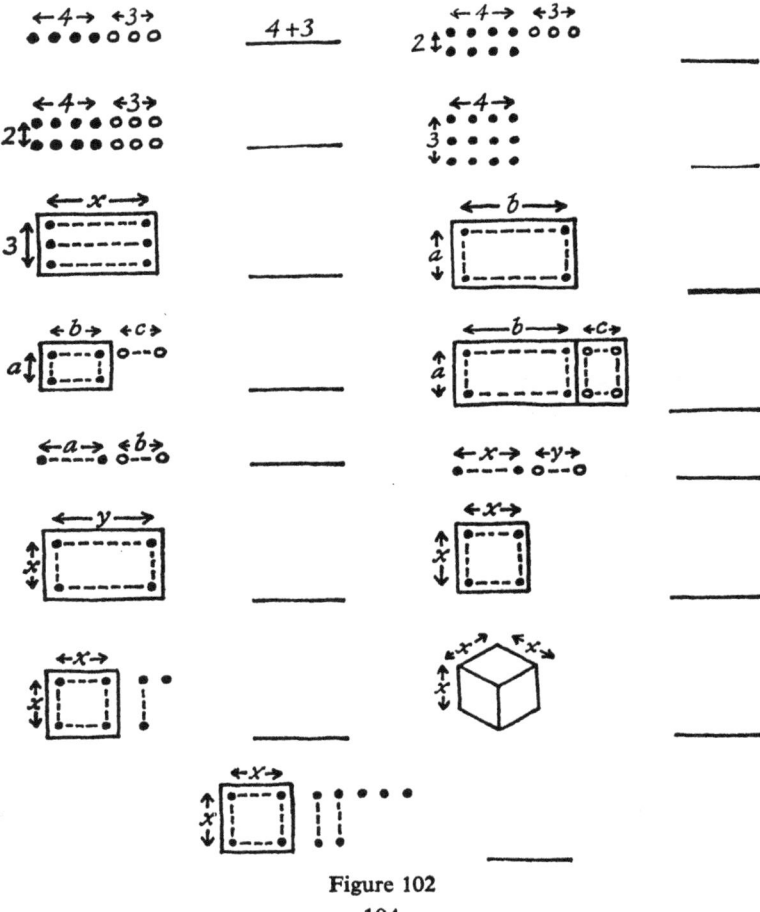

Figure 102

194

$$2x^5+ \ 3x^4+ \ 4x^3+ \ 5x^2+ \ 6x+ \ 7$$
$$8x^5+ \ 9x^4+10x^3+11x^2+12x+13$$
$$\overline{10x^5+12x^4+40x^3+16x^2+18x+20}$$

and will understand that similar principles are involved here.

It is interesting that the difficulty in teaching addition is not to get the student to do something but to get him to refrain from doing something. Children do not find it hard to see that 2 cubes and 4 cubes make 6 cubes, and that 5 squares and 6 squares make 11 squares. The difficulty, in an addition exercise such as that mentioned earlier, is to stop the children, when they have reached the correct answer $6x^3+11x^2$, from spoiling it all by going on to write $17x^5$ or some similar nonsense.

One can illustrate the meaning of $3x^3+7x^2$ by considering a particular value of x. Suppose x, 'the number thought of', were 2. Then x^3 would be 8 and x^2 would be 4, so $3x^3+7x^2$ would correspond to '3 eights and 7 fours'. Here is a picture of 3 eights and 7 fours (Figure 103).

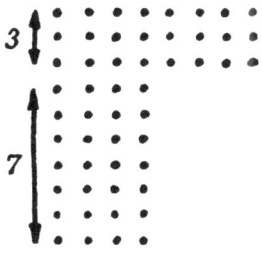

Figure 103

Now 3 and 7 make 10, but this is not a picture of 10 eights, nor is it a picture of 10 fours, nor in any simple way of 10 anythings. There are, in fact, 52 dots in the picture, and 52 is not a multiple of 10.

SUBTRACTION

At this stage, there is little to be said about subtraction. A person who understands addition understands subtraction. 'Subtract

3 from 5' is the same problem as '3 and what make 5?' 'Subtract $2a+3b$ from $7a+6b$', is the same problem as 'What do you have to add to $2a+3b$ to get $7a+6b$?'

Some teachers deal with addition and subtraction in algebra by interpreting a as 'apples' and b as 'bananas'. To explain $(7a+6b)-(2a+3b)$ they would say, 'If you took 2 apples and 3 bananas away from 7 apples and 6 bananas, what would you have left? It would be 5 apples and 3 bananas, so the answer is $5a+3b$.' Now $5a+3b$ is certainly the correct answer, so that this procedure does give a way of getting the right answer. This explanation has been under fire from mathematicians, so it seems worth while to examine its virtues and faults.

The first objection is that it obscures the fact that, in algebra, a and b stand for *numbers*, not for fruit!

The second objection follows from this. There are certain things we can say about numbers that we cannot say about apples and bananas. The equation $a=2b$ makes perfectly good sense in algebra: 'The number Alf thought of is twice the number Betty thought of.' But there is no sense in which we can say that 1 apple is 2 bananas. So a child brought up on the apple-banana approach might be very puzzled by an equation such as $a=2b$.

What are the virtues, if any, of the apple-banana idea? Teachers use it, of course, because it provides some definite image which children can think about. The teacher's instinct here is sound – to provide some object, within the children's experience, which they can visualize and reason about. But, of course, if the image is a distorted one, the ease of reasoning will be bought at a high price.

Just how distorted then is it? Surely it cannot be an accident that this method always yields the correct answer! The explanation seems to be the following. Suppose we change our approach just a little, and say that a is to be the *weight of an apple in ounces*. Then a does stand for a number. And yet at the same time a is associated with an apple. Now suppose we put together 5 apples and 2 apples to make 7 apples. There will be a corresponding relation between their weights. The weight in ounces of 5 apples added to the weight in ounces of 2 apples will equal the weight in ounces of 7 apples. So we may correctly write $5a+2a=7a$. But we still are able to cope with an equation such as $a = 2b$, if we are told that an apple *weighs as much* as 2 bananas.

In a matter of this kind, the mathematicians usually feel that the teachers are inaccurate, and the teachers feel that the mathematicians are pedantic. Every analogy is imperfect. An analogy is justified when the attention of pupils is directed towards the sound part of it, and when they are aware of its limitations. In Chapter 3, my motive for having sons climbing on to their fathers' heads was much the same as that which causes teachers to talk about apples – the belief that children think easily about something they can see. I tried to stress, however, that each man was *m feet high*, so that *m* stood for a number, not for a man. If, in teaching, we continually make remarks such as 'What *number* are you going to choose for *m*?' the children should not forget that the letters stand for numbers. The pictures will help the pupils to think, and not lead them into errors.

We can, of course, visualize $7a + 6b$ in terms of numbers of objects, as in Figure 104.

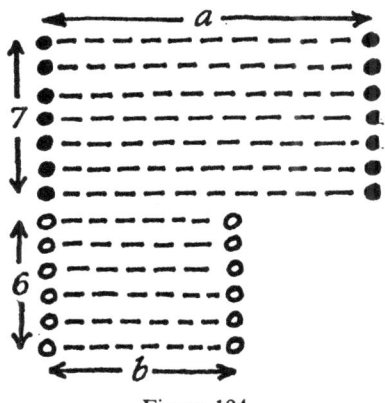

Figure 104

It is then fairly easy to see what would remain if we took away $2a + 3b$.

Much the same remarks apply to subtracting $2x^2 + 3x + 4$ from $9x^2 + 8x + 7$. The operation itself is not difficult:

$$9x^2+8x+7$$

Subtract $\quad 2x^2+3x+4$

$$\overline{7x^2+5x+3}$$

The pupil can see that the last two rows add together to make the top one. We can also use our picture with squares, lines, and dots. If from 9 squares, 8 lines, and 7 dots you remove 2 squares, 3 lines, and 4 dots, what remains? Here again our pictures are intended to suggest *numbers*; a line contains x dots, a square contains x rows with x dots in each.

MULTIPLICATION

An intelligent little girl, who had been taught arithmetic by rote, was liable to work out 13×12 by the following system:

$$\begin{array}{r} 13 \\ 12 \\ \hline 26 \\ 13 \\ \hline 273 \end{array}$$

The answer is clearly incorrect – for one thing, it gives an odd number for 13 dozen, in spite of the fact that a dozen is an even number. The answer also is much too big. 13 twenties make only 260; the girl's answer for 13 twelves is larger than 260, which is strange. But, for a child working from memory unaided by vision, the mistake is understandable. The work above does not look much different from the correct procedure:

$$\begin{array}{r} 13 \\ 12 \\ \hline 26 \\ 13 \\ \hline 156 \end{array}$$

In one method, the work slopes to the right, in the other, it slopes to the left – surely a small difference!

A child can be sure of avoiding such errors if he pictures what he is doing in multiplication. We picture 12×13 by a rectangle

198

with 12 rows of 13 dots. Why is 13 so written? Because it contains 1 ten and 3 units. Why is 12 so written? Because it contains 1 ten and 2 units. All of these features are shown in Figure 105.

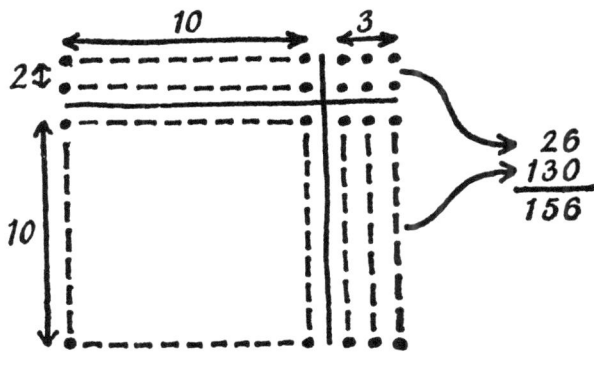

Figure 105

The 26 shows how many dots we have in the upper part of the rectangle; 2 tens and 6 units. The 130 shows what we have in the lower part; 1 hundred (a square 10 by 10), 3 tens, and 0 unit. Naturally, we add 26 and 130 to find how many dots there are in the entire rectangle.

There is nothing special about *ten*. If, in Figure 105, we were to replace 10 by any number x, we should have a diagram to show that $x+2$ rows with $x+3$ dots in them contain 1 square (with x^2 dots in it), 5 lines (with x dots in each), and 6 dots. In short, this picture shows that for any whole number x, we have:

$$(x+2)(x+3)=x^2+5x+6$$

Thus the picture we have found useful in arithmetic can be carried straight over to algebra. Indeed, multiplication in algebra is rather easier than in arithmetic. In arithmetic, if we get 17 tens we have to express this as 1 hundred and 7 tens. But in algebra, if we meet $17x$, we leave it as it is, for we do not know what number x represents. Thus we do not know how many x may be exchanged for one x^2; it is, therefore, impossible to make any transfers between the x column and the x^2 column. For example,

$(x+8)(x+9)=x^2+17x+72$. One can easily see (from a picture) that this result is *correct*. There remains the question; is it the simplest, neatest, shortest possible way of writing the answer?*

It is in fact; although we have 72 units, there is no way we can cut this number down, for we do not know the rate of exchange; how many units for one x.

This is one more example of an answer in algebra that to someone accustomed only to arithmetic looks unfinished. The difficulty here is not to teach the pupils how to perform some operation, but to persuade them to stay their hands – to convince them that the work is already complete.

The picture we used for the multiplication 12×13 can be adapted for more advanced multiplications such as 112×123. We imagine 112 rows with 123 dots in each, forming a rectangle. We remind ourselves of the meaning of 112, that it is 1 hundred, 1 ten, and 2 units, and similarly we recall the meaning of 123. In Figure 106, the rows and columns are broken up in such a way as to emphasize the meanings of 112 and 123.

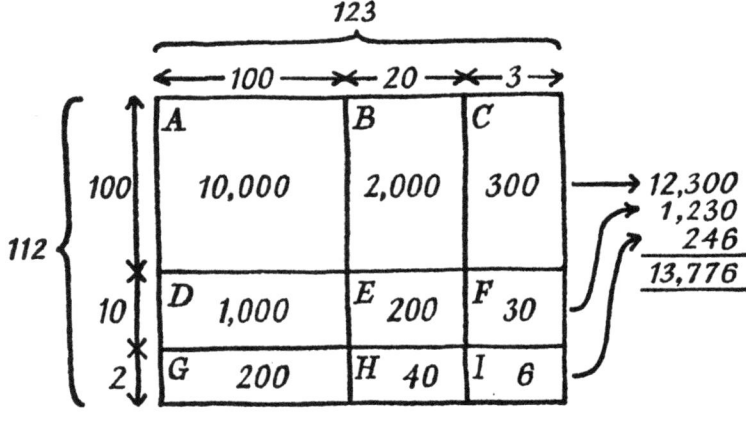

Figure 106

* In teaching it is important to make this distinction between work that is *wrong*, that makes *untrue* statements, and work that is in some sense incomplete – for example, if an answer were given as 17/34, and the pupil failed to observe that this *correct* answer could be written *more simply* as $\frac{1}{2}$.

The Routines of Algebra I

The whole rectangle, containing 112×123 dots, is thus broken into the regions marked A, B, C, D, E, F, G, H, I. Within each region has been written the number of dots it contains. To find the number in the whole rectangle, we must add together the numbers in the various regions. This calculation is shown on the right. The first number, 12,300, shows how many dots are in regions A, B, C; the second number 1,230, is for the dots in regions D, E, F; the third number, 246, is for the dots in regions G, H, I. Adding these we get 13,776 for the number of dots in the entire rectangle; that is, 13,776 is the value of 112×123.

It will be seen that the numbers written on the right of Figure 106 are those that would be written in one of the standard ways of performing the multiplication 112×123. Some people would reverse the order; they would write 246 at the top, then 1,230 and finally 12,300. This merely means that they would note first of all how many dots occur in regions G, H, I; then those in D, E, F; finally those in A, B, C. This is clearly just as good a procedure as the one we have followed. It is purely a matter of taste in what order one adds these numbers.

Thus Figure 106 shows that, without being told any rules, it is quite possible for a person to think out the procedure for multiplication, provided he understands the idea of illustrating multiplication by means of rectangles.

There are, however, some points that call for further comment. As 100 is 10×10 or 10^2, it would be natural to represent 100 as a square containing 100 dots. In this way we would bring out the fact that 100 is 10×10. However, when we come to 100×100, which is the number of dots in region A, we should have difficulty in representing this multiplication, if each 100 was pictured in the form of a square. The difficulty was mentioned in Chapter 1. It is often impossible to find a picture that will show at one glance all the information we would like it to. We have to sacrifice something. In Figure 106, in order to show 100×100 as the dots in region A, we have had to sacrifice the important aspect that $100 = 10 \times 10$. We write 100 along the top and side of region A, and we simply have to remember that 100 is 10 tens. The picture does not bring this fact out, it does not present it to our eyes. Each row of A consists of 100 dots arranged in line, and, of course, it is perfectly correct to imagine 100 dots in line. We can arrange 100 dots to form a square, but we do not have to.

In region A, we have written 10,000 for the number of dots it contains. Now this will not be obvious to all pupils. The question, 'What is a 100 hundreds?' seems to be quite a controversial one. Children will give answers ranging from 'A thousand' to 'A million'. So this question deserves analysis.

We need to go back to the basic ideas of the writing of numbers, discussed in Chapter 2. 100 is explained as being 10 tens, 1,000 as 10 hundreds, and 10,000, of course, is 10 thousands. We may write:

$$100 = 10 \times 10$$
$$1,000 = 10 \times 100 = 10 \times 10 \times 10$$
$$10,000 = 10 \times 1,000 = 10 \times 10 \times 10 \times 10$$

If we express everything in terms of 10, we can arrive at the answer for a 100 hundreds. For $100 = 10 \times 10$. So 100 times 100 $= 10 \times 10$ times $10 \times 10 = 10 \times 10 \times 10 \times 10 = 10,000$, from the table above.

Or we may put it as follows. 100 hundreds is 10×10 hundreds, which is 10 thousands (since 10 hundreds $= 1,000$).

All of these are, so to speak, written arguments. They do not convey a picture to the mind. They may satisfy children who think easily in abstract terms, but be of little help to children who have not yet reached this stage. So we consider some ways of picturing 100 hundreds.

Figure 107 shows 100 little squares. Suppose each square contains 100 dots. Then we have in the figure 100 times 100 dots. How many is this in all? If we take each row, it contains 10 hundreds – that is, 1,000 dots. We have 10 rows, so the whole figure contains 10 thousand dots.

One might also use the tree diagram, explained in Chapter 5. It is a little cumbrous to draw a tree with 10,000 twigs. Perhaps a description of a family tree will serve the purpose. John Smith had 10 sons, and his family continued the tradition – each son had 10 sons, and so it continued for several generations. Thus John Smith had 100 grandsons, 1,000 great-grandsons, and 10,000 great-great-grandsons (see Figure 108).

Each of John Smith's grandsons was thus himself a grandfather of 100 boys. So John Smith had 100 grandsons, and each of these, 2 generations later, was the ancestor of 100 boys. So, 4 generations after John Smith, the number of boys must have been

Figure 107

100×100. But we agreed earlier that John Smith had 10,000 great-great-grandsons. So 100×100 and 10,000 must mean the same thing.

What we have just done with 100×100 is a particular example of something we shall meet again many times in different forms. There is nothing special about the number 10. Our problem, to

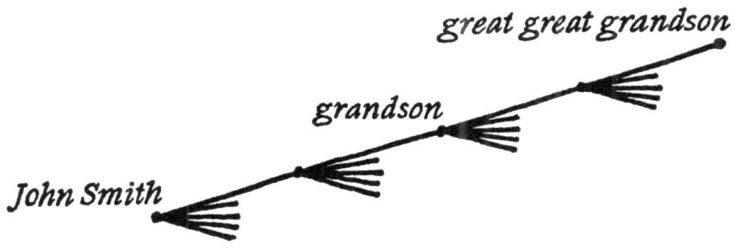

Figure 108

find 10×10 times 10×10 could equally well have been to find 7×7 times 7×7 or $x \cdot x$ times $x \cdot x$.

In Chapter 5, we noticed that repeated multiplication could be done in many ways. For example, if we had to calculate $2 \times 3 \times 4$

203

×5 we could begin with 2, multiply by 3 to get 6, then multiply by 4 to get 24, and finally by 5 to get 120. Equally well, we could work in the reverse order – begin with 5, multiply by 4 to get 20, then by 3 to get 60, and finally by 2 to get 120. Again, we could break this expression into two parts, 2×3 and 4×5. We could find 2×3 to be 6 and 4×5 to be 20 and then calculate $6 \times 20 = 120$. All of these, we observe, lead to the same answer. One might think there would be less scope for this kind of thing when the product contained the same number repeated, as in $10 \times 10 \times 10 \times 10$, instead of 4 different factors, as in $2 \times 3 \times 4 \times 5$. But there is still some room for variation. The most natural way, I suppose, to multiply out $10 \times 10 \times 10 \times 10$ would be to start with 10, find 10 tens to be 100, then 10 hundreds to be 1,000, and 10 times 1,000 gives the final answer, 10,000. But we could also do as we did when we split $2 \times 3 \times 4 \times 5$ into 2×3 and 4×5. We could say that $10 \times 10 \times 10 \times 10$ is 10×10 multiplied by 10×10, that is to say, 100 times 100. So 100×100 must give the same answer as we found by the other method, namely 10,000.

Another example, to show that repeated multiplication by the same number gives scope for the variety, is $2 \times 2 \times 2 \times 2 \times 2 \times 2$. Here again, the most natural way to calculate this product would be to begin with 2 and keep doubling; we should find in turn the numbers 2, 4, 8, 16, 32, 64. But we could also break it up into the parts $2 \times 2 \times 2$ and $2 \times 2 \times 2$, and get 8×8. Or we could break it into 2×2 and $2 \times 2 \times 2 \times 2$ and get 4×16. Each method, of course, gives the same final answer, 64.

To bring out the significance of these results, we want to write them so that it is clear that all the numbers involved, 4, 8, 16, 64, arise from repeated multiplication by 2. But we already know a shorthand suited for this purpose. Using indices, as explained in Chapter 5, we have the results:

$$64 = 2 \times 2 \times 2 \times 2 \times 2 \times 2 = 2^6 \text{ for short}$$
$$16 = 2 \times 2 \times 2 \times 2 \qquad\quad = 2^4$$
$$8 = 2 \times 2 \times 2 \qquad\qquad\; = 2^3$$
$$4 = 2 \times 2 \qquad\qquad\qquad\; = 2^2$$

We can now write our result $8 \times 8 = 64$ as $2^3 \times 2^3 = 2^6$ and the result $4 \times 16 = 64$ as $2^2 \times 2^4 = 2^6$.

Numbers which are obtained if we start with 2 and keep multiplying by 2 are often referred to as *powers of* 2. Here is a table of the first 12 powers of 2.

$2^1 = 2$	$2^5 = 32$	$2^9 = 512$
$2^2 = 4$	$2^6 = 64$	$2^{10} = 1,024$
$2^3 = 8$	$2^7 = 128$	$2^{11} = 2,048$
$2^4 = 16$	$2^8 = 256$	$2^{12} = 4,096$

Incidentally, many young children who enjoy arithmetic spontaneously make such a table. They start with 2 and keep doubling. This activity seems to intrigue them in some way.

An experiment with the above table is to choose two numbers in it (not too close to the end of the table) and multiply them together. Suppose, for example, we choose 32 and 64. We find $32 \times 64 = 2,048$. Ask the pupils if they notice anything about the result. They will probably observe that 2,048 is in the table above. In fact, our multiplication may be written $2^5 \times 2^6 = 2^{11}$. Do not tell the children any rule. Let them experiment, see what they observe, and, when they are thoroughly familiar with the results, let them try to account for what happens. They are almost certain, after translating $8 \times 16 = 128$ into the form $2^3 \times 2^4 = 2^7$, to notice that the indices 3, 4, and 7 have something to do with addition. They may find an explanation; for example, that $2^7 = 2 \times 2 \times 2 \times 2 \times 2 \times 2 \times 2$ can be split into $2 \times 2 \times 2$ times $2 \times 2 \times 2 \times 2$, which is 8×16. This throws light on why addition of indices is involved; of the 7 factors in 2^7, 3 are absorbed by 8 and the remaining 4 by 16. It is also possible to explain the result in terms of a family tree or of population. If the population of a country doubles in every generation, in 3 generations it will be doubled 3 times (that is, it is multiplied by 8); in 4 more generations, it will be doubled 4 times (that is, multiplied by $2 \times 2 \times 2 \times 2$ or 16); in the whole period of 7 generations, it has been doubled 7 times (that is, multiplied by 128).

After some experimenting for themselves, children begin to use the addition of indices as a short cut. They may work out $2^3 \times 2^5$ by writing it in full, $(2 \times 2 \times 2) \times (2 \times 2 \times 2 \times 2 \times 2)$, and seeing that this is 2^8, for it contains the factor 2 eight times. But when it comes to $2^{13} \times 2^{15}$ they will feel that life is too short to write out all these 'twos', and they will arrive at 2^{28} by using $13+15$ $=28$. They spot the pattern in the simpler cases, and use it to avoid work in the harder questions.

The numbers 10, 100, 1,000 . . . that we use in our usual system of writing numbers are, of course, the powers of 10. We may write $100=10^2$, $1,000=10^3$, and so on. Our result 100×100 $=10,000$ may be written $10^2 \times 10^2=10^4$ and agrees with our observations above on indices.

INVESTIGATION

In Chapter 2, we discussed the 7-finger system, the 3-finger system, and other systems for writing numbers. In the 3-finger system, how would $100_3 \times 100_3$ be written? What would be the answer to $100_7 \times 100_7$ as written in the 7-finger system? What happens in the other systems?

We have become involved in a considerable discussion, as a result of trying to justify why we wrote 10,000 in region A which contains 100 rows of 100 dots.

Fortunately, this discussion covers the main principle needed for filling in the remaining regions, B, C, D . . . and also for multiplication in algebra.

For example, in region B we want to write 100×20. As 100 $=10 \times 10$ and $20=2 \times 10$, this means we want to write $10 \times 10 \times 2$ $\times 10$. We have seen that multiplications may be carried out in any order, so we are entitled to write this as $2 \times 10 \times 10 \times 10$, which equals $2 \times 1,000$ or 2,000.

As was mentioned earlier, there is nothing special about the number 10. We could choose any number instead of 10 and use it to construct a picture similar to Figure 106. If, for example, we chose 7, we would replace 20, which is 2×10, by 2×7. Instead of 100, which is 10×10 or 10^2, we should use 7×7 or 7^2. However, this particular arithmetical example is not very interesting or useful. It might be used to explain the way multiplication is performed in the 7-finger system. It has no other evident purpose. So we may as well go directly to algebra, and suppose we replace 10 by x. Instead of 20, which is 2×10, we shall write $2x$. Instead of 100, which is 10×10 or 10^2, we shall write $x.x$ or x^2. Thus we shall be considering a rectangle with x^2+2x+3 dots in each row, and there will be x^2+x+2 rows. The picture will now be as shown in Figure 109.

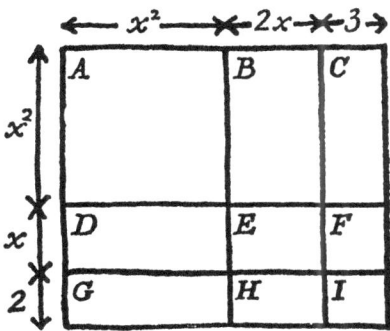

Figure 109

We now have to write in each region the number of dots it contains. The number in region A is x^2 times x^2. This means $x.x$ times $x.x$, which is $x.x.x.x$ or x^4. So we write x^4 in region A. (This argument runs along exactly the same lines as our first way of finding 100 times 100, on page 202.)

In region B, we have x^2 rows with $2x$ dots in each. So the number of dots in region B is x^2 times $2x$, which is $x.x$ times $2.x$ or $x.x.2.x$. As order of multiplication does not matter, we can write this as $2.x.x.x$, which is $2x^3$. So $2x^3$ is written in region B.

In region C we have x^2 times 3 dots. This is $x.x.3$, which is the same as $3.x.x$ or $3x^2$.

In D, we have x times x^2, which is x times $x.x$, which is $x.x.x$ or x^3.

In E we have x times $2.x$, which is $x.2.x$ or $2.x.x$ or $2x^2$.

In F we have $3x$; in G, $2x^2$; in H, $4x$; and in I, 6.

After writing these in, our picture appears as in Figure 110.

What then is the total number of dots in the rectangle? We have to add the numbers in the various regions. Here the cautions given on page 191 in connexion with addition need to be repeated. Any answer which lumps together the whole thing into a single term such as $24x^{18}$ is most certainly wrong. If necessary we can recall our pictures for x^3, x^2, and x. In B we have 2 cubes and in D 1 cube. So we have 3 cubes in all, representing $3x^3$. In C, E, and G we have squares, 7 in all, representing $7x^2$. In F and H we have

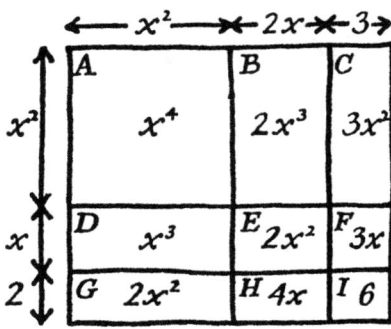

Figure 110

'long thin lines', 7 again, representing $7x$. We do not have any picture for x^4, but we hope that enough has been said to remind the pupils that x, x^2, x^3, and x^4 are different, and that we have to keep a separate account for each of them. The total cannot be written more simply than as $x^4+3x^3+7x^2+7x+6$.

It is perfectly satisfactory, when children are first learning to multiply in algebra, for them to draw pictures such as Figure 110 and then put together the squares, cubes, and so forth.

Later, they may be ready to set the work out more formally, like this:

Regions	Number
A, B, C	$x^4+2x^3+3x^2$
D, E, F	x^3+2x^2+3x
G, H, I	$2x^2+4x+6$
Total	$x^4+3x^3+7x^2+7x+6$

The total given above is the answer to this multiplication question.

If you look back to Figure 106 you will see how closely the multiplication in algebra follows the pattern of the multiplication in arithmetic.

The Routines of Algebra I

To do multiplication correctly a pupil must understand three things: (i) the general plan of campaign, as described above; (ii) how to find the number in each region; (iii) how to add these numbers.

Of these, (i) is fairly simple. If the pupil is familiar with the idea of representing multiplication by means of rectangles, there is very little to explain. Children see easily that the number in the whole rectangle can easily be found, if we can work out the number in each region.

The final step, (iii), the process of addition, is often incorrectly performed. This, however, has already been discussed in the section of this chapter under the heading 'Addition'.

There remains only step (ii). The most complicated situation that could arise would be if, for example, some region contained $4x^3$ rows with $5x^2$ dots in each. We should then need to find $4x^3$ times $5x^2$. This means $4.x.x.x$ times $5.x.x$, which is $4.x.x.x$ $.5.x.x$. The order of multiplication makes no difference, so we may write this as $4.5.x.x.x.x.x$, and this is $20x^5$.

In a little while, children will get tired of writing this out in full, and will begin to devise their own short cuts. They should be encouraged to do this, *provided their answers are correct*. But any time a wrong answer comes, the pupil should be asked to reason the question out by the primitive method indicated above. This makes sure that the children do not pick up some 'rule' that Tommy Smith's father recommends, which is a distorted version of some correct procedure.

At some stage in algebra (often much too soon) children are asked to perform calculations such as $2a^2b + 3ab^2$ times $4a^2b + 5ab^2$. The general plan of campaign is exactly the same. We draw a rectangle to represent the product, and break it into regions (Figure 111).

Region A now contains $2a^2b$ rows with $4a^2b$ dots in each. So we want to find $2a^2b$ times $4a^2b$. This means $2.a.a.b$ times $4.a.a.b$ or $2.a.a.b.4.a.a.b$. This is rather like the expressions that arise in arithmetic when a number is being broken into its factors, as, for example, in Chapter 5 we found $72 = 3 \times 2 \times 3 \times 2 \times 2$

and we rearranged this as $2 \times 2 \times 2 \times 3 \times 3$ or $2^3 \times 3^2$. We brought the factors 2 together and the factors 3 together, so that we could use indices and write our answer in a brief form. In algebra we change the order so that the factors a come together, and the factors b come together. So we rewrite $2.a.a.b.4.a.a.b$ as $2.4.a.a.a.a.b.b$ or $8a^4b^2$.

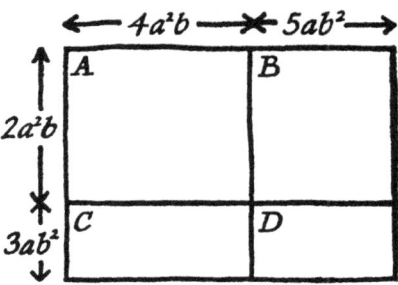

Figure 111

No new idea is involved here – only the same principle, that order of multiplication makes no difference. This idea is applied to a slightly more complicated situation. In this way we find the following; region A contains $8a^4b^2$ dots; B has $10a^3b^3$; C has $12a^3b^3$; D has $15a^2b^4$.

Now the problem of addition arises again. B contains 10 times the number a^3b^3; C contains 12 times the number a^3b^3. We can certainly combine these as 22 times the number a^3b^3, or $22a^3b^3$. The total may certainly be written $8a^4b^2+22a^3b^3+15a^2b^4$. Can we carry this any further? The answer is No. If you think of a as any number chosen by Alf and b any number chosen by Betty, we can expect a^4b^2, a^3b^3, and a^2b^4 to be three different numbers. If Alf chose 5 and Betty 2, these numbers would be 2,500; 1,000; 400. There is no obvious way of working out in your head the sum of 8 times 2,500, 22 times 1,000 and 15 times 400, other than actually carrying through the routine arithmetic. Of course, Alf and Betty *might* both choose the same number, say 2, in which case a^4b^2 and a^3b^3 and a^2b^4 would all mean 64. Then we could say, 8 sixty-fours and 22 sixty-fours and 15 sixty-fours make 45 sixty-fours, which would be a simpler form of the answer. Betty and Alf *might* do this, but we have no guarantee that they will. The whole point of the kind of algebraic calculation we are making at present is that it should be like the 'Think-of-a-Number' tricks; *whatever numbers are chosen*, we want our statements to be true. So we must not base our arguments on particularly simple cases that *might* arise.

In general, it is wise to work with multiplications in which only one letter is involved – such as x^2+2x+3 times $4x^2+5x+6$ – until the pupils have become thoroughly familiar with this type of multiplication, and have acquired some feeling for what is correct and what is incorrect. The more complicated type of multiplication, involving such things as a^4b^2 or $7x^5y^3z^8$, may be left until later. Perhaps such expressions will arise in the course of a problem, and can then be explained, when the need for dealing with them has been felt.

There is one very simple piece of multiplication involving several letters that may be introduced early. In Chapter 8 we were at some pains to emphasize that $(a+b)^2$ is not a^2+b^2 but a^2+b^2+2ab. This may suggest to some pupil the question, what would $(a+b+c)^2$ be? Surely not $a^2+b^2+c^2$; but what would it be? We can illustrate the problem by drawing a picture for $(2+3+4)^2$, as in Figure 112.

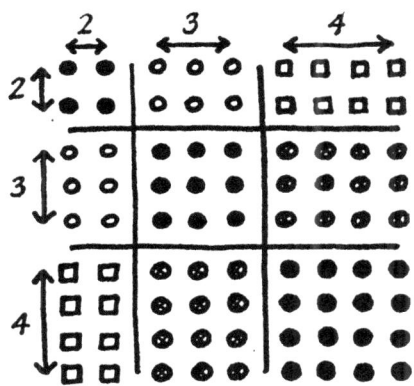

Figure 112

Here we see $2+3+4$ rows with $2+3+4$ objects in each row. The objects have been drawn in different ways, so as to emphasize

211

the regions into which the square is divided. The black dots form 3 squares. The number of dots in these 3 squares is $2^2+3^2+4^2$. The hollow circles fill 2 rectangles, each 2 by 3. The small, square objects fill two rectangles each 2 by 4. Finally, the shaded circles fill 2 rectangles, each 3 by 4. The total number of objects can be found by adding together the numbers in the various regions. In arithmetic, this procedure does not serve any obvious purpose. In algebra, however, this way of dividing up the square allows us to multiply out $(a+b+c)^2$. Figure 113 shows $a+b+c$ rows with

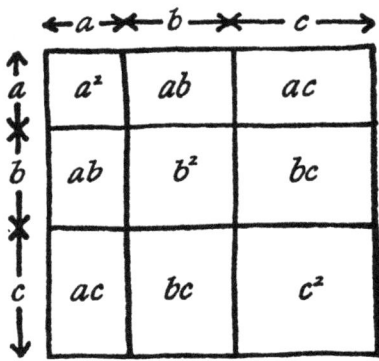

Figure 113

$a+b+c$ objects in each. In each region, we write the number of objects in it. Now we have 3 squares which together contain $a^2+b^2+c^2$ objects. We see two rectangles a by b, two rectangles a by c, and two rectangles b by c. The total number of objects inside the square is thus $a^2+b^2+c^2+2ab+2ac+2bc$.

In teaching, one reason for bringing in this work with $(a+b+c)^2$ is that it gives us an excuse for reminding the pupils of the result for $(a+b)^2$. If they have forgotten that $(a+b)^2$ is a^2+b^2+2ab, we get them to draw a picture resembling Figure 66 and to work this result out for themselves. When they come to think out $(a+b+c)^2$, and find that the answer contains not only $a^2+b^2+c^2$ but also $2ab$ and $2ac$ and $2bc$, this should still further remind them that $(a+b)^2$ contains $2ab$ as well as a^2+b^2.

Investigation. Draw similar pictures for $(a+b+c+d)^2$ and $(a+b+c+d+e)^2$ and multiply these expressions out.

212

The Routines of Algebra I

DIVISION

In the past, long division in arithmetic was commonly taught as a procedure that had to be memorized. Few people knew why they were doing what they did. Recently there has been some attention paid (particularly in the U.S.A.) to a way of doing division that allows us to think our way to this result.

Consider the instruction, 'Divide 532 by 7'. In what sort of situation might such an operation be necessary? We might have 532 objects which were to be shared fairly between 7 children. I will consider how this problem might be attacked if I did not know the 7 times table beyond the result $5 \times 7 = 35$! I might think, 'I can certainly give each child 50 objects.' This would use 350 objects, and leave 182. I would then see that I could give 20 more to each child. This would use 140 and leave 42. The calculation so far might be recorded as follows:

Number given to each child		
	532	Original number
50	350	
	182	remain after first hand-out
20	140	
	42	remain after second hand-out

I now see that I can give out 5 more to each child. This uses 35, and leaves me with 7. Now obviously I can give 1 object to each child, and finish up my store. So the complete record would appear like this:

	532
50	350
	182
20	140
	42
5	35
	7
1	7
	0

Each child has received altogether $50+20+5+1$, that is, 76 objects. So 532 divided by 7 is 76, and there is no remainder.

If my knowledge of the 7 times table had been even less, I might have used even more handouts, and yet arrived at the end at the correct answer. On the other hand, with experience of this method and a proper knowledge of the tables, I could go directly to the most efficient argument.

Number given to each child	532
70	490
	42
6	42
	0

Each child has received $70+6$ or 76 objects. I have found the same answer, and the work above is essentially *what we do in the standard procedure for long division.* So this approach allows children to think their way towards the usual procedure. As their knowledge improves, they make the work shorter and shorter. Yet the story behind the procedure allows them to know all the time what they are doing.

To check the correctness of the above division, we would multiply 76 by 7 and get 532. Therefore, a child who understands how to multiply should never hand in an incorrect answer to a division problem. The answer should be checked by multiplication. If an error has been made, it will be detected, and can be corrected. Suppose at some stage an error of copying had been made and the answer recorded not as 76 but as 67. On checking by multiplication, we should find that to give each of 7 children 67 objects required not 532 but only 469. We have a surplus of 63. This means we can give each child 9 more, and each child will end up with $67+9$ or 76 articles. Thus this story approach allows us to correct mistaken answers.

In algebra also we can picture division. Division is the opposite of multiplication; '12 divided by 3' is another way of asking 'What you must multiply 3 by to get 12?' or 'How many rows with 3 dots in each are needed to make 12 dots?' In the same way, to divide x^2+5x+6 by $x+2$ we have to think how many rows with $x+2$ objects in each would be needed to make x^2+

$5x+6$ objects. Needless to say, no pupil should attempt such a question unless he is thoroughly familiar with multiplication in algebra. Indeed, the best time for proposing such a division question is immediately after a pupil has completed several multiplications of the type given on page 180.

The problem can be posed pictorially. Here is a picture of x^2+5x+6 (Figure 114).

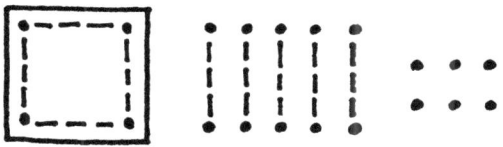

Figure 114

How many rows of $x+2$ dots can we make from this material? Since the square, representing x^2, contains x rows of x dots, it seems reasonable that we first try to make x rows of $x+2$ dots. This would use up the material shown in Figure 115.

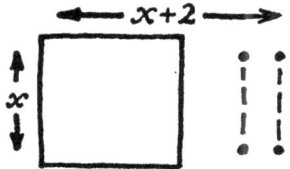

Figure 115

This leaves us with 3 'long thin pieces' and 6 dots, and it is not hard to see that this is just right for 3 rows of $x+2$ dots. So we end up with the arrangement in Figure 116.

Here again, making the rows uses up our supply of material in much the same way that handing out the objects to the children used up our initial supply. We can record how the supply dwindles.

Initial amount	x^2+5x+6
x rows of $x+2$ dots uses up	x^2+2x
and leaves	$3x+6$
3 rows of $x+2$ dots uses	$3x+2$
and leaves nothing	

Division is multiplication in reverse. Earlier in this chapter we multiplied x^2+2x+3 by x^2+x+2 and found $x^4+3x^3+7x^2+7x+6$ as the answer. We can use this result to compose a problem in division: 'Divide $x^4+3x^3+7x^2+7x+6$ by x^2+2x+3.' We can still use Figure 110 as a diagram, to illustrate the

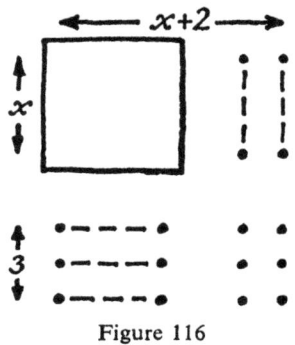

Figure 116

division. The question is, 'How many rows of x^2+2x+3 dots are needed to make a rectangle containing $x^4+3x^3+7x^2+7x+6$ dots? The answer actually is x^2+x+2, but we would not know that at this stage. So we must suppose x^2, x, and 2 erased on the left of the diagram. Nor would we yet know that x^4 is to be written in region A, $2x^3$ in region B, and so forth. These would appear in the course of the division. However, it is quite useful for us to have this multiplication before our eyes as we think out how to perform division, for it allows us to see a clue that could be used by someone who did not know the answer in advance. If you look at Figure 110, you will notice that x^4 appears in only one place – in region A. Figure 117 shows what someone would know if he was just starting to perform this division and was also given the clue that region A contains x^4 dots. He would be able to see that the question mark must be replaced by x^2, for x^2 rows of x^2 dots makes x^4 dots, and nothing else will do this.

Now fortunately, in any multiplication, the power of x that occurs in region A occurs nowhere else, and it is the highest power of x that occurs in the product. For example, if we multiply x^3+5x^2+7x+1 by x^4+2x^2+3x+8, region A will contain x^4

216

rows with x^3 dots in each, making x^7 in all. You will not find x^7 in any other region, nor will you find anywhere a higher power, such as x^8 or x^9. If you complete this multiplication, you will see that this is so, and it is not hard to see why it happens. By 'the highest power' we mean that the factor x is to occur as often as

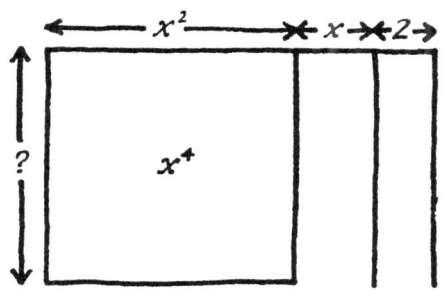

Figure 117

possible. For example, $2x^5$ contains a higher power of x than $73x^2$, for $2x^5 = 2.x.x.x.x.x$ contains the factor x five times, while $73x^2 = 73.x.x$ contains it only twice.

If you look at Figure 110 again, you will see that $x^2 + 2x + 3$ is broken into the pieces* x^2, $2x$, and 3, while $x^2 + x + 2$ is broken into x^2, x, and 2. The number of dots in each region is obtained by multiplying a piece of $x^2 + 2x + 3$ by a piece of $x^2 + x + 2$. For example, to get the number of dots in region E, we multiply the piece $2x$ by the piece x. Now naturally, if we want this number to contain as many factors x as possible, we should choose each piece to contain as many factors x as possible. In $x^2 + 2x + 3$ the piece with most factors x is x^2. In $x^2 + x + 1$ the piece with most factors x is also x^2. In region A we accordingly have the number of dots $x^2.x^2 = x.x.x.x = x^4$. There is no way of getting more than 4 factors, and there is no other way of getting as many.

In all this, I am supposing that the two expressions to be multiplied are written in the most usual way with the highest power of x coming first. For example, $x^2 + 2x + 3$ is written in that order and not as, say, $2x + 3 + x^2$. If this is done, then region

---

* The technical name for a *piece* is a *term*.

A, the first region in the first row, will always be the one in which the highest power of x is written.

We return to our division problem. How many rows of x^2+2x+3 dots are needed to make a rectangle containing $x^4+3x^3+7x^2+7x+6$ dots?

Our argument so far has used only the first term of each expression. In any question, 'How many rows of $x^2+\ldots$ are needed to make a rectangle containing $x^4+\ldots$ dots?' we could use the same argument. Region A must contain x^4 dots. To get x^4 dots in region A we must begin by having x^2 rows, as in Figure 118.

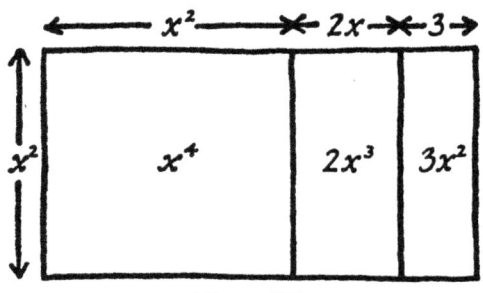

Figure 118

Making these x^2 rows uses up $x^4+2x^3+3x^2$ dots. So our statement looks like this:

Initial amount	$x^4+3x^3+7x^2+7x+6$
x^2 rows of x^2+2x+3 dots use	$x^4+2x^3+3x^2$
and leave	x^3+4x^2+7x+6

So now we have the question, how many rows of x^2+2x+3 dots can we make from the x^3+4x^2+4x+6 dots that are left. Now this is exactly *the same type of question* that we began with, but *simpler*. For now our highest power is only x^3 instead of x^4.

As before, we look only at the first terms. How many rows of $x^2+\ldots$ dots are needed to make $x^3+\ldots$ dots? To get x^3 we have to multiply x^2 by x. So we ought to make x more rows.

You can follow this on Figure 110. By making x^2 rows we filled regions A, B, C. By making x rows we fill regions D, E, F. So our account continues.

The Routines of Algebra I

$$\begin{array}{r}
\text{We had left} \quad x^3+4x^2+7x+6 \\
x \text{ rows with } x^2+2x+3 \text{ dots use} \quad \underline{x^3+2x^2+3x} \\
\text{and leave} \quad 2x^2+4x+6
\end{array}$$

We now have to see how many rows of x^2+2x+3 dots we can make from the $2x^2+4x+6$ that are left. It is fairly clear that 2 is the answer, but if we did not see this, we would have to look at the first terms; how many rows of $x^2+ \ldots$ to make $2x^2+ \ldots$? Answer, 2.

These last 2 rows fill regions G, H, I of Figure 110, and exactly use up our supply of dots.

It may, of course, happen on occasion that we cannot use all our dots. If, as in the last example, we were making rows of x^2+2x+3 dots, it might happen (in some other problem) that, at a certain stage, we had made a number of rows and had $4x+1$ dots left over. We cannot make another row from these. They simply have to be left as a remainder.

Exercises

1. Draw a picture of x^2+x+1 dots. Rearrange the dots you have drawn so that there as many rows of x dots as possible. How many rows would there be? How many dots would be left over? What then is the result of dividing x^2+x+1 by x?

2. Draw a picture of x^2+5x+6 dots. How many rows of $x+3$ dots could you make out of the dots you have drawn? How many dots would be left over?

3. Draw a picture of $2x+7$ dots. How many rows of $x+2$ dots could you make from these, and what would be left over?

4. Draw a picture of x^2+2x+1 dots. How many rows of $x+1$ dots could you make from these?

5. Draw a picture of x^2+2x+5 dots. How many rows of $x+2$ dots could you make from these? How many dots would you have left over?

6. Divide x^2+3x+2 by $x+3$.

7. Saying '23 divided by 7 gives 3 with remainder 2' is the same as saying '23 equals 3 sevens and 2'. This last statement may be written $23=3 \times 7+2$. Write similar equations to express:
   (i)   That 9 divided by 2 gives 4 and remainder 1

(ii)   That 36 divided by 10 gives 3 and remainder 6

(iii) That 1,057 divided by 100 gives 10 and remainder 57

8. The statement '$x^2+5x+13$ divided by $x+2$ gives $x+3$ and remainder 7' says the same thing as the equation $x^2+5x+13 =(x+3)(x+2)+7$. Write the equations that express the following statements:

   (i) $2x+5$ divided by $x+1$ gives 2 and remainder 3

   (ii) x^2+x+1 divided by x gives $x+1$ and remainder 1

   (iii) x^4+x^2+x+1 divided by x^2+1 gives x^2 and remainder $x+1$

9. Divide $x^2+5x+20$ by $x+3$. Write an equation, as in question (8), to express your result. Check your work by testing whether this equation is correct.

10. Find $(x^2+x+1)^2$. That is, multiply x^2+x+1 by x^2+x+1. Also work out 111^2.

Compare your answers. Why do they involve the same numbers?

11. Find $(x^2+2x+2)^2$ and 122^2. Compare your answers.

12. Find $(x^2+2x+3)^2$ and 123^2. Why do we not get here the similarity that we observed in questions (10) and (11)?

13. Find (i) $(2x+1)^2$

          (ii) $(2x^2+2x)^2$

          (iii) $(2x^2+2x+1)^2$

Show that adding the answers to (i) and (ii) together gives the answer to (iii).

If a triangle had sides of length $2x+1$, $2x^2+2x$ and $2x^2+2x+1$, what kind of triangle would it be? (You may get a hint if you choose some particular number for x, such as 1 or 2. You can then draw the triangle and see what you notice. Any-

Figure 119

one familiar with geometry will know a method for drawing a triangle with sides of specified lengths. But, even without compasses, it is easy to make such triangles. If I wanted to see what a triangle with sides 6, 7, and 8 looked like, I could cut three sticks, 6 inches, 7 inches, and 8 inches long, and put them together to make a triangle (see Figure 119).

COINCIDENCES IN NUMBER BASE SYSTEMS

With our ordinary way of writing numbers we have the results $11 \times 11 = 121$, $12 \times 12 = 144$, and $11 \times 12 = 132$. Some of the exercises at the end of Chapter 2 draw attention to the existence of very similar results with bases other than 10. For example, in base 3 the result $4 \times 4 = 16$ takes the form $11_3 \times 11_3 = 121_3$, which looks very much like the result for 11 elevens. Similarly we find $11_4 \times 11_4 = 121_4$, $11_5 \times 11_5 = 121_5$, and so on. In fact it looks very much as if the result $11_n \times 11_n = 121_n$ works for any number n from 3 upwards. (We cannot consider it in binary, because 0 and 1 are the only digits permitted in binary.)

In our ordinary system, 121 means 1 hundred, 2 tens, and 1 unit. 100 of course is 10 tens or 10^2. So $11 \times 11 = 121$ written in full means:

$$(10+1)(10+1) = 10^2 + 2 \times 10 + 1$$

The result $11_7 \times 11_7 = 121_3$ similarly means:

$$(7+1)(7+1) = 7^2 + 2 \times 7 + 1$$

In fact these are instances of the result:

$$(n+1)(n+1) = n^2 + 2n + 1$$

which, as we have seen, holds for any number n.
In the same way, the results of the form $12_n \times 12_n = 144_n$ are another way of expressing the multiplication:

$$(n+2)(n+2) = n^2 + 4n + 4$$

So also, results such as $11_7 \times 12_7 = 132_7$ and $11_4 \times 12_4 = 132_4$ are particular cases of the general result:

$$(n+1)(n+2) = n^2 + 3n + 2$$

221

Vision in Elementary Mathematics

EVEN AND ODD

In Chapter 1, we considered the properties of even and odd numbers, and their properties were demonstrated by drawing pictures. It was mentioned that these properties could be shown by means of formal algebra. This we will now consider.

One mistake has to be avoided at the outset. Suppose we consider the simplest problem of all; show that even added to even gives even. A pupil may produce the following: 'An even number is of the form $2n$. So even plus even is $2n+2n=4n$. But $4n$ is sure to be an even number. So the sum is even. Q.E.D.'

The trouble here is that too much has been 'proved'. For $4n$ is not merely an even number. It is divisible by 4. Now it is *not* true that the sum of any two numbers is divisible by 4. For example, 6 and 8 are even but $6+8=14$, which is not a multiple of 4.

Where is this 'proof' faulty? It lies in using $2n+2n$. Admittedly $2n$ represents any even number, but $2n+2n$ represents any even number *added to itself*. For example if n, the number chosen, is 3, then $2n+2n$ means $6+6$. If we choose 4 for n, then $2n+2n$ means $8+8$. There is no way of choosing $2n+2n$ to make this mean $6+8$.

In $6+8$, the 6 is twice 3 and the 8 is twice 4. Two different numbers, 3 and 4, are involved. Two different letters must be used to represent them.

We might organize the random choice of two even numbers in the following way. Tell Alf to think of a whole number. We use a for the number chosen by Alf. Double Alf's number. This will certainly give us an even number, $2a$ for short. Now tell Betty to think of a whole number. We double that and get our second even number, which is twice Betty's number, or $2b$ for short.

The result of adding the two numbers together is $2a+2b$.

We can see that $2a+2b$ is what you get if you double $a+b$. Now $a+b$ is a whole number, so $2a+2b$ is twice a whole number, that is, it is even. This does prove what we wanted to show.

We can see our particular example $6+8$ in it. If Alf thought of 3 and Betty of 4, $2a+2b$ would mean $6+8$. According to the algebra, this sum should be twice $a+b$, which is in this case twice $3+4$. As 14 is twice 7, everything is as it should be.

This point of two numbers being involved arises in every question of this type. To prove that odd times odd is odd, we have to multiply $2a+1$ by $2b+1$ and make sure that the result is odd. We find $(2a+1)(2b+1)=4ab+2a+2b+1$. Now $4ab+2a+2b$ is twice $2ab+a+b$, hence even. The answer to our multiplication is 1 more than this, so it must be odd.

At the end of Chapter 1, the question was raised of what happened when a number of the *type* $7n+4$ was multiplied by one of the *type* $7n+6$. We get Alf to help us choose the first number and Betty the second. So the first number will be $7a+4$ and the second $7b+6$. If we multiply these together we find:

$$(7a+4)(7b+6)=49ab+42a+28b+24$$

This may look a complicated answer, but most of it is quickly dealt with. $49ab$ is exactly divisible by 7; so are $42a$ and $28b$. So a remainder on division by 7 can only arise from the last number 24, which in fact leaves a remainder of 3. So the answer to the multiplication is of the type $7n+3$.

CHAPTER TEN

The Routines of Algebra – II

IF you compare Chapter 9, which we have just left, with a typical algebra textbook, one point should strike you. Almost from the beginning, most algebra textbooks have problems containing many minus signs. There are a few minus signs in Chapter 9, but you have to search for them. There is a definite reason for this. More mistakes in algebra are due to a misunderstanding or misuse of minus signs than to any other single cause. It therefore seems reasonable to delay the handling of minus signs as long as possible. We do not, of course, want our pupils to think that minus signs are mysterious and incomprehensible. We can think about minus signs and see their meaning just as clearly as with any other topic in algebra. But they certainly do lend themselves to careless little mistakes. Before attacking minus signs it is, therefore, desirable for a pupil to have had as much experience as possible of the rest of algebra; to have seen what algebra is about and to have acquired some feeling for the subject; to have formed the habit of checking every statement carefully; to have worked many problems correctly and built up confidence; to have formed the habit of independent thinking; by attacking obstinate problems, to have become patient and persistent. If an algebra book in its earliest chapters introduces several fiddling rules for operating with minus signs, these rules are likely to affect the whole taste of the subject. The pupil may have insufficient experience to see why these rules are right and reasonable. Their early introduction may thus block the most important thing of all in beginning algebra – the child's realization that this is something he can think out for himself.

In this book an effort has been made, by using such devices as boys standing on men, bags of stones, and squares filled with dots, and also by keeping most of the problems simple, to make children feel that an appreciable part of algebra is straightforward and almost obvious. In this way, it is hoped, confidence and interest can be fostered. Even so, we have on several occasions touched on the question of minus signs. Figure 16 on page 55 illustrates the problem of subtracting $m-s$ from $m+2s$. On pages 83–4 of Chapter 4, we discussed Nancy Cochrane's discovery of how to subtract $x-1$ from $2x$. Note, incidentally, that every single child in a class of very able youngsters answered this question wrong at the first attempt. So such a question is highly unsuitable for early or hasty treatment, particularly with classes of average children. In Chapter 7, in order to deal with the Bears Problem, we were led to study $50-(20-x)$ and $50-(20+x)$. We have, so to speak, skirted round the territory of the minus sign. We have used minus signs whenever a problem forced us to do so, or when we found an easy question involving them. This manner of approaching a difficulty is, I believe, sound teaching practice. It gives pupils a chance to become familiar with a topic without involving them too quickly in questions of complicated detail.

It is probably worth repeating that the correct use of minus signs does not present any great intellectual difficulty; rather it is a matter of attention to irksome detail. We have already mentioned one such detail; children easily fall into the error of reading $a-b$ as 'take a from b'. This is typical of the slips made with minus signs. Usually there is a choice between two things. Is the answer $x+7$ or $x-7$? Do we mean $a-b$ or $b-a$? Should we add or subtract? The decisions are small ones, but usually there are several such decisions to be made, and one small mistake will mean a completely wrong answer at the end.

One reason why many children make these slips is *that nothing depends on the result*. There is all the difference in the world between a calculation in real life and in the school room. In school, a slip means a cross on the paper, and no one is much the worse for it. In life, a slip may mean the collapse of a building, the bankruptcy of a business, or the ruin of an individual. The consciousness of these consequences forces us to check the details of our work. In an unreal situation, such as exists in a typical classroom, no such force operates. A few children have the type of mind that is ready to check details patiently. A few more are doing well at the subject, they are enjoying it, and have some pride in their success; they pay attention to detail in order to continue having this satisfaction. The rest are essentially not trying. Whatever they may believe in their conscious minds about mathematics being necessary to worldly success, something deep within them feels that this is a waste of time. They try to learn just enough to keep out of trouble. They have a hazy recollection of some rule; they have not troubled to find out exactly what the rule is, nor exactly to what circumstances it applies. Inevitably, they obtain wrong answers. It is quite impossible for a pupil to do good work while in this state of mind. We may prevent it by good teaching at a critical stage, so the pupil goes from success to success and comes to regard algebra as one of his 'good subjects'. We may relate mathematics to the pupil's own interests and enthusiasms. We may perhaps bring into the classroom some of the reality of adult life, by getting children to design something which is actually going to be made. We may interest children by giving them the experiences of discovery and insight.

Suppose the emotional situation to be satisfactory – the pupil is reasonably confident and interested in the subject. How is he to see what ought to be done with minus signs? We consider the standard processes in turn.

ADDITION

Many of the simpler problems in addition can be visualized directly.

Example 1. Add $x-2$ to $x-3$. As in Chapter 4, we visualize $x-2$ as 'a bag from which 2 stones have been removed' and $x-3$ as 'a bag with 3 stones removed'. Adding corresponds to putting these together, as shown below.

We see 5 holes through which stones have been removed. Our picture represents 2 bags, from which 5 stones have been taken away. Our answer is $2x-5$. We may write:

$$\begin{array}{r} x-2 \\ \text{Add} \quad x-3 \\ \hline 2x-5 \end{array}$$

Example 2. Add $x+2$ to $x-3$. We picture $x+2$ as a bag and 2 stones; $x-3$ is pictured as in the previous example. The picture below shows the sum of $x+2$ and $x-3$.

What are we to do about this? The children themselves, with little or no prompting, will suggest that we use the 2 stones on the left to replace 2 of the stones missing from the bag on the right. We demonstrate this on our picture, by erasing the 2 stones and 2 of the holes. We are left with 2 bags and a single hole, representing the total $2x-1$.

$$\begin{array}{r} x+2 \\ \text{Add} \quad x-3 \\ \text{Total} \quad \overline{2x-1} \end{array}$$

Example 3. Add $x-2$ to $x+3$. The picture will now be as below.

With the 3 stones on the right, we can fill both the gaps in the left hand bag, and still have a stone left. We finish then with 2 bags and a single stone; the answer is $2x+1$.

$$\begin{array}{r} x-2 \\ x+3 \\ \hline 2x+1 \end{array}$$

227

Example 4. Add $x+2$ to $x+3$. This offers no difficulties. We have already answered questions of this type in Chapter 4. We get 2 bags and 5 stones, or $2x+5$ as our answer.

You may notice that the above 4 examples cover all the possibilities for this type of question. Some textbooks follow the perfectly horrible procedure of labelling these Case (1), Case (2), Case (3), Case (4), and giving a rule for each. For example (3) above the rule would read something like this: 'When we are dealing with 1 plus sign and 1 minus sign, and the number with the minus sign is larger than the number with the plus sign, the answer will contain a minus sign, and the number in the answer will be the difference of the numbers in the addends.'

There is no need to remember this jargon, nor even to understand what it says. I mention it only to point out that it is much easier to see the answer, to think it out for yourself, than either to understand or to remember the rule.

Example 5. Add $x+2c$ and $x-3c$. We may vary our picture a little. Instead of a bag, we will represent x by a pail (of stones, or grains of sand, or other objects). Of course, x stands for the *number* of objects in the pail. We think of c as the number of objects required to fill a cup. We may then picture $x+2c$ as a pail with two cupfuls added, and $x-3c$ as a pail with two cupfuls removed. Figure 120 shows these put together. The arrows indicate that we next use the 2 extra cupfuls in the first pail to fill the 2 hollows in the second. We are left with 2 pails, from which 1 cupful has been removed. The answer is $2x-c$.

$$x+2c$$
$$\underline{x-3c}$$
$$2x-c$$

If a child works these 5 examples on the same sheet of paper, he will probably be struck by the resemblance between Example (5) and Example (2). The questions are similar; so are the answers. On the basis of this resemblance, many children will be able to guess, correctly, the answers to further questions. For example, adding $x-2c$ to $x+3c$ recalls Example (3). We guess that the answer will be $2x+c$, and indeed it is. Children might also be able to guess what would be the effect of altering the numbers involved. For instance, without drawing pictures, they might well be able to guess the result of adding $x-2c$ and $x+6c$.

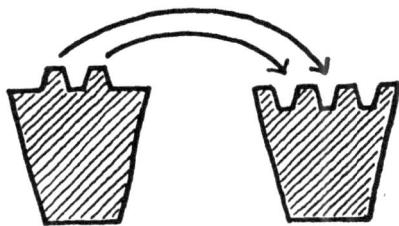

Figure 120

THE TRANSITION TO FORMAL WORK

It is, of course, an essential part of learning algebra that, at a certain stage, one ceases to think in pictures and begins to go by the look of the symbols written on the page. This, I suppose, is the solitary correct idea behind the learning of rules. Someone who works regularly with algebra does come to be rather like a calculating machine. If he sees $2x$ and $3x$, he knows immediately that their sum is $5x$. Some mechanism in his brain automatically produces this response. The mistake in the 'rule' approach is trying to go too directly to this final state.

If we teach children algebra by means of pictures, we will find that the children themselves begin to invent their own formal short-cuts. For example, suppose that, with the help of bags and stones, we teach them to carry out a calculation such as:

$$\text{Add} \quad \begin{array}{r} 3x+4 \\ 5x+6 \\ \hline 8x+10 \end{array}$$

We give the class a dozen exercises on this theme. In a short time, most of the children will have noticed that the work breaks up into two separate addition questions. In the first column we have $3+5=8$, in the second $4+6=10$. The children will, in fact, work their exercises in this way. They will not go to the trouble of drawing bags or even imagining them.

One could, of course, justify their procedure in terms of the picture. We combine 3 bags and 5 bags to make 8 bags, also 4 stones and 6 stones to make 10 stones. Two quite separate additions are involved. But I very much doubt if children think it out in this way. I suspect that, after working the first two or three questions out by means of pictures, they observe that the answers could be obtained by performing two additions, and to save themselves time and labour, they do the remaining questions that way. It is not the deductive reasoning of an adult mathematician. It is a much looser kind of thinking: 'We noticed this; we thought it would work; it did.'

However large the element of guesswork in this approach, it yet works much better than the memorizing of rules. The children have arrived at the method by their own experience and their own observation. Often when children are told a rule, they do not hear it correctly, or they misunderstood what it said. But with our approach, the message is not conveyed in words at all. The children observe for themselves that *doing* certain things saves a lot of work and leads to the same answer as the longer method of drawing pictures. If a child gets out of practice, and forgets the details of the method, he has only to work one or two questions the long way round – by drawing pictures – and in a minute he will make the same observations again, and will have re-discovered the quick, formal method. Thus as compared with rote memorization, the system of allowing children to discover and devise their own rules is more reliable and more permanent.

When minus signs are present, the same kind of separation is possible. Consider again Example (5), the addition of $x+2c$ and $x-3c$. In Figure 120 on page 229, each x is illustrated by the number of objects required to fill a pail. The effect of $2c$ is shown by 2 mounds sticking up, and of $-3c$ by 3 hollows. We could vary the problem; with $x+5c$ and $x-8c$ there would be 5 mounds and 8 hollows, but there would still be 2 pails. If we wish to add $x+ \ldots c$ and $x- \ldots c$, where any numbers whatever may be written in the spaces, we shall still have a picture with 2 pails in it. So we can be sure that the answer will begin with $2x$.

Again, if we look at the c part of the question, we have $+2c$ and $-3c$. These will be represented by 2 mounds and 3 hollows.

It does not matter what they are mounds on or hollows in. We can use 2 mounds to fill 2 hollows, and we shall be left with 1 hollow, which represents $-c$, the removal of 1 cupful.

Thus our addition question again breaks into two parts.

$$\begin{array}{ccc} x+2c & x & +2c \\ \underline{x-3c} & \underline{x} & \underline{-3c} \\ 2x-\ c & 2x & -\ c \end{array}$$

The first addition here breaks up into the two smaller problems. We have by now long been familiar with the addition of x and x to make $2x$. The addition of $+2c$ and $-3c$ to make $-c$ is novel. We can adapt the ideas of Figure 120 to picture it. We start with a level beach; $+2c$ indicates that 2 cupsful of sand are to be added; $-3c$ that 3 cupsful are to be removed. Figure 121 illustrates the situation. Tipping the 2 mounds into 2 hollows, we are left with 1 hollow, or $-c$.

Figure 121

The line of thought we have followed is quite a natural one. We added $x+2c$ and $x-3c$ with the help of a picture. This surely was not a difficult problem. Then we noticed that our work fell into two parts, (i) seeing how many pails the picture contained, (ii) seeing how the mounds and hollows affected the answer. Our pictures enabled us to split the original question into two parts, and to deal with these. The two parts were (i) how to combine x and x, (ii) how to combine $+2c$ and $-3c$.

A textbook writer always faces the problem whether to follow the logical or the psychological order. The psychological order is from complex to simple. We start with a problem such as that of the men and their sons in Chapter 3. Several different processes may be needed to solve the problem, but at least the purpose of the work is clear. After solving several such complicated

231

problems we gradually become aware that the calculations again and again involve certain simple operations, such as combining $+2c$ and $-3c$.

The logical order, which many textbooks follow, reverses the discovery order. We start with simple, standard operations, such as combining $+2c$ and $-3c$. We work many exercises on these, and are supposed to become proficient in handling them. At the end (if we survive this method of instruction) we may find that, by combining the various operations we have learnt, we can solve some significant and interesting problems.

It is easy to see why the logical approach appeals to writers. It seems so reasonable to build up proficiency in each detail and then to combine these skills. In the same way, there are certain games which it may be wise not to start playing straight away, but rather to do some exercises which ensure that the correct way of performing some action becomes habitual. Otherwise the player may be saddled for life with some incorrect habit which he is unable to unlearn. But in such a situation, the player at least has seen the game played, and knows the purpose of the exercises he is doing. Now in mathematics this is not so. Few people, glancing at the first dozen pages of an algebra book, could form any idea at all of what algebra eventually enables us to do. The exercises appear totally meaningless. For most human beings it is depressing and frightening to be subjected to instruction in which you have no idea what you are dealing with, where you are going, or why you are going there. The resulting atmosphere of fear inhibits all thought. The learner abandons all effort to grasp the true nature of what he is studying, and clutches at rules and hints, which are meant to be a substitute for understanding.

At the very least, then, a person embarking on the study of algebra is entitled to a preview of the subject, to some kind of indication of the problems it enables us to solve, and of the way in which the solution of complicated problems can be broken down into a series of small, routine operations. These operations thus come to have an interest for us which otherwise they would not have.

The preview is a minimum demand. More satisfactory is to give the student some samples of problems in algebra which he can solve with the minimum of technical preparation. Having had the satisfaction of solving these, and finding that there are other problems which he cannot solve because his technical knowledge is insufficient, he has then both the background and the motive for a more systematic study of the subject. The order of the chapters in this book is an attempt to realize this ideal. But this book makes no claim to be the last word on the subject. Much more work remains to be done, in making a collection of problems that would serve as an introduction to algebra, and would be sufficiently varied to offer something to readers of all ages and tastes.

A POSSIBLE SOURCE OF CONFUSION

In adding $x+2c$ and $x-3c$ we were led to discuss how we should combine $+2c$ and $-3c$. Since the original problem was one of addition, instead of asking how to *combine* these terms, we may ask how to *add* them. Now here is a tangling of associations. We are performing addition, and yet the pupil sees the minus sign in $-3c$, which he associates with subtraction. No wonder that at this stage the pupil who is only vaguely aware of what is happening anyhow begins to get confused.

In this context, we are still using adding in its natural sense of *putting together*. If two businesses merge into one, they do not merely combine the assets they had before; the new firm is also responsible for all former debts. If we add together £2c and a *debt* of £3c, we end with a *debt* of £c. This is a hackneyed illustration of the result that 2c added to $-3c$ gives $-c$.

This style of thinking is already used in elementary arithmetic. If a child is mentally adding 1,000,002 and 999,997, he may check the correctness of his answer by reflecting that the first number is 2 more than a million, while the second is 3 less than a million. The sum should be 1 less than 2 million. Here, he is quite naturally and unconsciously adding 2 *more* and 3 *less*. It does not strike him that he is adding $+2$ and -3. However, the numbers are $1,000,000+2$ and $1,000,000-3$, and the result is thus a particular case of $x+2$ added to $x-3$.

233

One may also picture such additions in terms of arrows, as we did in Chapters 3 and 7. In Figure 122, we first see $x+2c$ combined with $x-3c$. In the second drawing, the order of the arrows

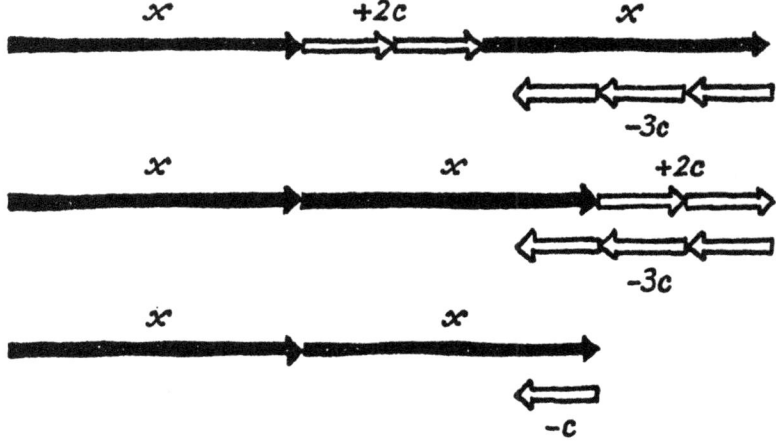

Figure 122

has been changed, but the final result is still the same. In the third drawing, we see that $2x-c$ gives a simpler way of specifying the final position.

SUBTRACTION

Figure 32 on page 95 showed four ways of imagining subtraction. A little earlier, three questions were given, each of which corresponded to the expression $5-3$. Needless to say, pupils should not embark on the techniques of subtraction in algebra if they are not thoroughly familiar with the meaning of subtraction.

If addition has been mastered, subtraction becomes much easier, since $a-b$ corresponds to the question 'What do we have to add to b to get a?'

Earlier we considered the addition of $+2c$ and $-3c$. We now consider the two subtractions associated with these expressions. What is $-3c$ subtracted from $+2c$? This is the same as saying, what must be added to $-3c$ to give $+2c$? The answer is $+5c$;

to 3 hollows you have to add 5 mounds if the final result is to be simply 2 mounds.

The other way round, what is $+2c$ from $-3c$? What do you have to combine with 2 mounds to produce 3 hollows? The answer is 5 hollows, corresponding to $-5c$.

Various illustrations are possible. What must be combined with £2 to give a debt of £3? Answer, a debt of £5. What must happen to a temperature of $+2°$ to give a temperature of $-3°$? Answer, a *fall* in temperature of $5°$, which suggests the answer -5.

We may use the interpretation of $a-b$ that asks 'How much bigger is a than b?' This has variants – how much hotter is a degrees than b degrees? – how much better is it to have £a than £b? – how much taller is a man of a feet than one of b feet? We have to agree that a height of -5 feet means that a man's head is 5 feet below ground level!

With subtraction, there may always be an element of confusion as to which is being taken from which. This confusion is least when the subtraction is written in the vertical form:

$$\begin{array}{r} 10c \\ \text{subtract} \quad -2c \\ \hline 12c \end{array}$$

Here it is usually clear to pupils that the two bottom rows, $12c$ and $-2c$, should add to make the top row. Indeed the best way to perform this subtraction is to think, $-2c$ and what make $10c$?

If the subtraction is posed in a form such as 'Subtract -5 from -2', the pupil may well be uncertain whether the answer should be 3 or -3. While such uncertainty continues, it may be worth while for the pupil to write out at length a statement involving a familiar case, and carefully compare it. For example:

10 from 17 means '10 and what make 17?' -5 from -2 means '-5 and what make -2?' The pupil makes sure that -5 is written under 10 and -2 under 17 throughout. The answer in fact is 3 or $+3$ (these have the same meaning). One may use the same comparison to test whether the answer is reasonable.

10 from 17 is $+7$ because $17°$ is hotter than $10°$. -5 from -2 is $+3$ because $-2°$ is hotter than $-5°$. Here again there are the

usual variants – because a debt of £2 is preferable to a debt of £5; because 2 feet below sea level is higher than 5 feet below sea level.

The most confusing form of all is $(-2)-(-5)$. This still means '-5 from -2' but we now have an array of minus signs that dazzles the eye. It is essential not to rush. We observe that this subtraction is concerned with -2 and -5. We sketch part of a thermometer (Figure 123).

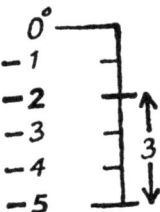

Figure 123

We note that $-2°$ and $-5°$ are $3°$ apart. So the difference between the temperatures is $3°$, and our answer must be either $+3$ or -3. Which is it? We reflect that $-2°$ is hotter than $-5°$, so we choose two familiar temperatures (such as $10°$ and $17°$ which we used earlier) and write the hotter temperature to correspond to $-2°$, and the colder one to correspond to $-5°$. We thus reach the comparison:

$$17-10$$
$$(-2)-(-5)$$

As $17-10$ is positive, so also should $(-2)-(-5)$ be positive.

Incidentally, we would be led to consider $(-2)-(-5)$ in the course of carrying out a subtraction such as $(x-2)-(x-5)$. This breaks up into two parts just like the additions considered earlier. In the first column we would subtract x from x and get 0. In the second column we would subtract -5 from -2, the operation we have just been considering. Now it is very easy to check the statement $(x-2)-(x-5)=3$. This says 'If from 2 less than any number, you subtract 5 less than that number, you get the answer 3.' This surely is true. If for example we choose the number 100, we find $98-95$ is 3, as predicted.

If, instead of $(-2)-(-5)$ we had been considering $(+2)$ $-(-5)$ our picture of the thermometer would have been as in Figure 124.

The two temperatures are now 7° apart. By any of the methods considered earlier we can check that the answer is $+7$, not -7.

Here we have the usual feature of 'either-or' choices. If we have a subtraction in which the figures 2 and 5 occur, together

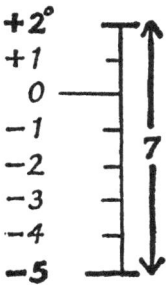

Figure 124

with assorted plus and minus signs, the answer will either involve 7 (which is $5+2$) or 3 (which is $5-2$). The answer will be either positive or negative. Thus there are four conceivable answers, $+7$, $+3$, -7, and -3. If you guess at random, your chance of getting the right answer is twenty-five per cent. If you learn rules by heart, your chance remains about the same. If you are ready to go slowly, master any linguistic conventions involved (such as that $5-3$ means 'from 5 take 3'), and make sketches to show the meaning of each question, you have every prospect of reaching 100 per cent proficiency.

Experienced mathematicians, when they are in a hurry or when their attention is directed elsewhere, continue to make slips in signs throughout their working lives. If you reach the stage where you are sure of getting the right answer when some vital issue depends on your doing so, provided you are given time to check your result, you may feel reasonably content.

'REMOVAL OF BRACKETS'

There is a rule taught in connexion with expressions such as $k-(a+b+c-p-q)$. This rule states that we may erase the brackets, provided that we change the sign of every term inside the bracket. That is, this rule tells us that the answer should be $k-a-b-c+p+q$. The letters that had plus signs inside the bracket now have minus signs; those that had minus signs now have plus signs. The first letter inside the bracket, a, has no sign attached to it, but it is understood that a and $+a$ have the same meaning.

This rule of course applies only when there is a minus sign in front of the bracket. That is to say, this rule is essentially a rule for carrying out subtraction. It is mentioned here because you may find references to it in textbooks. The justification for this rule can be seen by anyone who understands subtraction. Essentially, it is an extension of the ideas contained in results (i) and (ii) on page 155 and can be thought out by the same argument. Suppose I set out shopping with £k in my pocket; I buy articles costing £a, £b, and £c, but receive refunds of £p and £q. The amount I spend is thus £$(a+b+c-p-q)$ and my balance at the end of the day is found by subtracting this from £k. So it is $k-(a+b+c-p-q)$ pounds. But, by regarding the prices of the articles as money paid out and the refunds as money paid in, I could reach the answer $k-a-b-c+p+q$ pounds.

We could illustrate this result by means of arrows, by elaborating the type of diagram used in Figures 80 and 86.

The result could also be presented in terms of mounds and hollows. I have mounds containing a, b, and c stones, and hollows p and q stones have been scooped out. What must I add to this to get simply a mound of k stones? I must provide the mound of k stones, hollows of size a, b, and c (to wipe out the mounds of size a, b, c) and mounds of sizes p and q (to wipe out the hollows). The result is $k-a-b-c+p+q$.

It should be noted here that this result does not apply only to expressions containing single letters such as k, a, b, c, p, q. We could apply it to, for example, $x^3-(2x^2y+7x^2+5xy-9t^2-2)$ and find that this expression was equivalent to $x^3-2x^2y-7x^2-5xy+9t^2+2$.

Beginners in algebra often fail to realize that a symbol, such as k, may be replaced not only by 'any number we like', but also by some other algebraic expression. In the example above, k has been replaced by x^3, a by $2x^2y$, b by $7x^2$, and so on. Such a procedure merely elaborates the story. In the shopping narrative, I began with k. But where did I get that amount? Possibly some eccentric millionaire asked me to think of a number. I told him the number, he worked out the cube of that number and gave me the corresponding number of pounds. If x is short for the number I thought of, x^3 would be short for the number of pounds I started the day with. So we may with perfect propriety replace k by x^3.

If we prefer, we may run through the story itself with suitable amendments. 'I began the day with £x^3. I bought objects costing £$2x^2y$, £$7x^2$, and £$5xy$, on which I was given refunds of £$9t^2$ and £2. How much had I at the end of the day?' We reach the same result again.

MULTIPLICATION

Multiplication involving minus signs comes into arithmetic very early. Often children cannot remember 9×7. A child may work it out by thinking that 10 sevens make 70. If we remove 1 seven from 10 sevens we shall be left with 9 sevens. So $9 \times 7 = 70 - 7 = 63$. As $9 = 10 - 1$, here we are arguing that $(10-1) \times 7 = 10 \times 7 - 1 \times 7 = 70 - 7 = 63$.

Children use the same type of argument for 9×6. This is $60 - 6$ or 54.

Another troublesome multiplication is 8×7. We could argue that 8 sevens are 10 sevens with 2 sevens removed; thus the answer is $70 - 14$ or 56. Figure 125 illustrates this last argument. Each column contains 7 objects. We first make 10 columns and then cross 2 of them out. The picture shows $(10-2) \times 7 = 10 \times 7 - 2 \times 7$.

Exercise. Is there anything special about the numbers 10, 2, and 7? Could we draw a similar picture using, say, 9, 3, and 5? Could we use any numbers a, b, and c in place of 10, 2, and 7? (For the present, we had better suppose a to be larger than b.) What equation in algebra would this lead us to write?

The idea we have used above occurs often in small problems of arithmetic. What is the cost of 5 objects at 19s. 11d. each? If we presented a pound note for each object, we should receive a penny change on each. So the total cost must be 5d. less than £5, or £4 19s. 7d. What is 23×99? The same kind of argument shows that it is 23 less than 23×100, so the answer is 2,277.

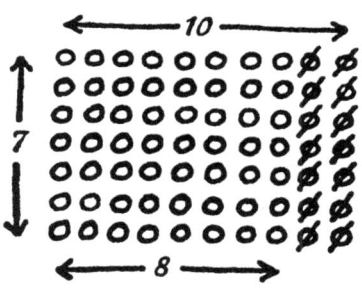

Figure 125

Here again there is continuity between arithmetic and algebra, and a formula such as $(a-b)c=ac-bc$, which at first sight appears utterly strange and unfamiliar, is found to be merely the written statement of a well-known idea.

EXPERIMENTS WITH SQUARES

In Chapter 8 we observed children arriving for themselves at the conclusion that $(x+1)^2$ is x^2+2x+1. It is very often necessary to deal with $(x-1)^2$. How could children investigate this question? We saw in our earlier investigation that if we wanted to enlarge a square 3 by 3 into one 4 by 4 we could begin by putting 2 threes on to it (as in A of Figure 126). We are then well on our way to having 4^2 objects. If we want to cut down from 3 by 3 to 2 by 2 (as at B in Figure 126) it might seem natural to remove 2 threes. In the same way, if we had x rows of x objects, and we wanted to cut it down to $(x-1)$ rows of $(x-1)$ objects, it might seem a natural first step to remove 2 sets of x objects. This is an idea only; it still has to be tested. If we subtract $2x$ from x^2 how does the result compare with $(x-1)^2$? Children could choose different numbers for x, and tabulate the results. The table below shows the square numbers, x^2, the numbers $2x$ to be subtracted, the results

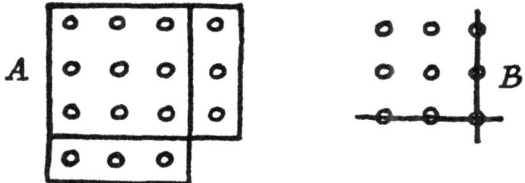

Figure 126

of the subtraction, x^2-2x, and $(x-1)^2$, the numbers we hope eventually to obtain.

x	2	3	4	5	6
x^2	4	9	16	25	36
$2x$	4	6	8	10	12
$x-2x$	0	3	8	15	24
$(x-1)^2$	1	4	9	16	25

It is now clear that x^2-2x is not the same as x^2, but it is pretty close. In fact, if we add 1 to each of the numbers in the x^2-2x row, we get the numbers in the $(x-1)^2$ row. This means $(x-1)^2=x^2-2x+1$.

We have now succeeded in multiplying out $(x-1)^2$, but we do not have a very clear picture. In fact it seems much harder to deal pictorially with $(x-1)^2$ than with $(x+1)^2$. In Figure 126 A above, after putting the two flaps on the square of 3, it is evident to the eye that just one more object is needed to make the square of 4. There is no difficulty in generalizing this picture and showing that by adding two flaps of n objects and a single object to the square of n we can make the square of $n+1$. In fact we had this figure on page 171. So far as I know, no one has devised an equally natural and convincing picture to illustrate the result $(n-1)^2 = n^2-2n+1$. In Figure 126 B, we have cut 3^2 down to 2^2, but it is by no means evident to the eye that we have crossed out 2 threes and added 1. At first sight it might even appear that we have simply removed 2 threes, for we have crossed out 1 row and

241

1 column. But this cannot be right for $9-2\times3=9-6=3$, and we can still see 4 objects not crossed out. In fact one object has been crossed out twice. We might amend Figure 126 B, to make it appear as in Figure 127.

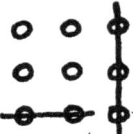

Figure 127

We now see that you can cut down from 3^2 to 2^2 by removing $3+2$ objects. If instead of 3^2 we had started with n^2, our first stroke would erase n objects, and the second stroke would erase $n-1$ objects. In all we thus remove $2n-1$ objects, and so leave $n^2-(2n-1)$ objects. It is good for pupils to think this result out this way. The more often they think a result out by different methods, the more likely they are to remember it. Also, a little exercise in subtraction occurs naturally in this work. They will see that it leads us once more to the answer n^2-2n+1. But they do have to use an *argument* to reach the result. This approach falls short of the ideal of direct vision. For an important result, we would like to have a diagram, which we could show the pupils any time we wanted to remind them of this result. They should immediately see from the diagram what the result is. Also, the diagram should follow naturally from some simple line of thought so that the pupils would be able to draw it for themselves whenever they were in doubt.

We can demonstrate the process of cutting 3^2 down to 2^2 by the series of actions shown in Figure 128.

Here we have removed 3 twice and have added 1. If we begin with n^2, a similar process would remove n twice, and add 1.

We can replace this sequence of actions by a single picture if we add the 1 at the beginning, as in Figure 129. From 3^2+1 we have erased 2 threes.

Children should draw similar pictures to show 4^2+1 with 2 fours erased, 5^2+1 with 2 fives erased, and so on, until the general moral is clear to them.

242

After this discussion the pupils themselves might well raise the question – what happens with $(n-2)^2$? They may first experiment with a particular case. Since $5-2=3$, we might consider cutting 5^2 down to 3^2 (Figure 130).

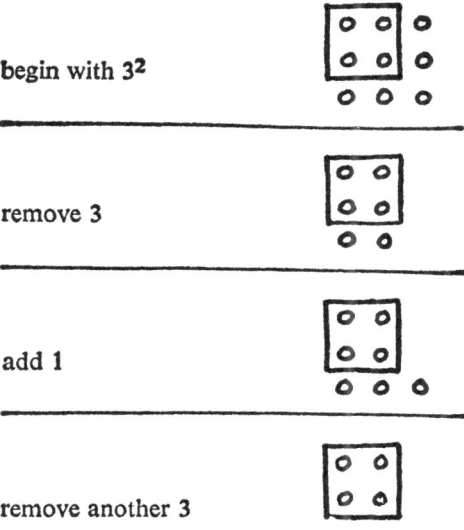

begin with 3^2

remove 3

add 1

remove another 3

Figure 128

We can now repeat all our earlier types of attack. We may have a rash first thought – we have crossed out 4 fives. But $25-4\times5$ $=25-20=5$ and we have 9 objects remaining. We want to find some excuse for adding 4, since $9=5+4$. Ah, yes; there are 4

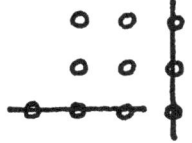

Figure 129

objects in the south-east corner that have been crossed out twice. So $3^2=5^2-4\times5+4$, and the generalization is suggested $(n-2)^2$ $=n^2-4n+4$.

We can justify this by an argument involving subtraction.

What we have actually removed from 5^2 is 2 fives and 2 threes. What we remove from n^2 is 2 columns of n objects, and then 2 shortened rows of $n-2$ objects. In all we remove $2n+2(n-2)$ $=4n-4$ objects. The final result is n^2-4n+4.

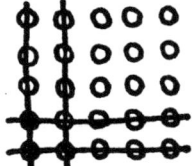

Figure 130

We can give a dynamic demonstration. Begin with 5^2; remove 2 columns of 5; add 4 objects, to restore the bottom 2 rows to their original length; remove 2 rows of 5.

Finally we can replace this sequence of actions by the static picture (Figure 131).

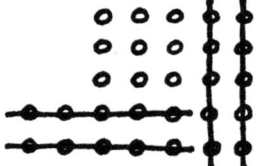

Figure 131

To 5^2 we have added 4 and then removed 4 fives. In general, to n^2 we should add 4 and then remove $4n$.

At this stage it may be convenient to tabulate our results, together with some found earlier in Chapter 8.

$$(n+4)^2 = n^2 + 8n + 16$$
$$(n+3)^2 = n^2 + 6n + 9$$
$$(n+2)^2 = n^2 + 4n + 4$$
$$(n+1)^2 = n^2 + 2n + 1$$
$$n^2 = n^2$$
$$(n-1)^2 = n^2 - 2n + 1$$
$$(n-2)^2 = n^2 - 4n + 4$$
$$(n-3)^2 = \ldots\ldots\ldots$$
$$(n-4)^2 = \ldots\ldots\ldots$$

244

The pupils can now probably guess the results for $(n-3)^2$ and $(n-4)^2$. It is good for them to guess these, for they will only guess correctly if they have seen the pattern of the answers. To see the pattern they have to examine the table carefully.

If we look down the first column on the right hand side, we see n^2 every time. It seems probable that n^2 continues to occur throughout this column. In the next column, there is a gap in the middle. It is convenient to imagine this gap filled by $0n$, for no n occurs in the result for n^2. The numbers in this column then run 8, 6, 4, 2, 0, -2, -4. This pattern is easy to continue with -6 and -8. We may argue either that these numbers fall by 2 at each step, or that, as we go away from the gap in the middle we get 2, 4, 6, 8 . . . with plus signs above the gap and with minus signs below.

The pattern in the last column is entirely different. We have the same numbers 1, 4, 9, 16 as we go away from the gap, and with the same sign, $+$. If we include 0 for the gap, the numbers run 16, 9, 4, 1, 0, 1, 4, 9, 16. At first this seems surprising. If we had temperatures of $16°$, $9°$, $4°$, $1°$, $0°$ on successive days, we might well think, 'The temperature is falling all the time. Soon it will be below zero.' But if we examine the figures a little more closely, we may begin to doubt this prophecy. While the temperature falls, it does so more and more reluctantly. From $16°$ to $9°$ is a fall of $7°$; from $9°$ to $4°$, a fall of $5°$; from $4°$ to $1°$, a fall of $3°$; and from $1°$ to $0°$ is a fall of only $1°$. It is not really so surprising that after this the temperature begins to rise. In fact if we write the numbers 16, 9, 4, 1, 0, 1, 4, 9, 16 together with the changes from one number to the next, we see that the process is a perfectly regular one.

Numbers	16		9		4		1		0		1		4		9		16
Changes		-7		-5		-3		-1		$+1$		$+3$		$+5$		$+7$	

The pattern of the table on page 244 can be summed up in the two statements below. (Compare question (4) on page 172.)

$$(n+a)^2 = n^2 + 2an + a^2$$
$$(n-a)^2 = n^2 - 2an + a^2$$

The first of these results is essentially the same as the result we had for $(x+y)^2$ on page 132, and can be illustrated by writing n for x and a for y in Figure 66. The second result is new.

We have followed one possible path to the discovery of this result. Pupils may hit on some other way. For example, in Chapter 8, we observed the table in Figure 132.

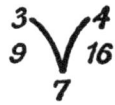

Figure 132

4^2 exceeds 3^2 by 7, which is $3+4$. One might apply this idea to finding $(n-1)^2$. For n^2 should exceed $(n-1)^2$ by $(n-1)+n$, which is $2n-1$. This means that $(n-1)^2$ should be $2n-1$ less than n^2, so $(n-1)^2=n^2-(2n-1)=n^2-2n+1$. Again, we may work back from $(n-1)^2$ to $(n-2)^2$. The difference between these should be $(n-1)+(n-2)=2n-3$. Subtracting $2n-3$ from n^2-2n+1, we get n^2-4n+4 for $(n-2)^2$. We can continue in this way to obtain $(n-3)^2$, $(n-4)^2$ and so on. This work provides exercises in subtraction in a natural way. This approach, of course, merely develops an idea already mentioned on page 242.

THE INVESTIGATION EXTENDED

The idea already used to find $(n-a)^2$, which is short for $(n-a)$ $(n-a)$, can also be used to investigate multiplication of the type $(n-a)(n-b)$.

Let us consider once again 7×8. As $7=10-3$ and $8=10-2$ we are considering the multiplication $(10-3)(10-2)$. If we began with 10 rows of 10 objects, to make this into 7 rows of 8 objects, we erase the bottom 3 rows and the last 2 columns, as in Figure 133.

Our first rash thought might be that we had removed 3 rows of 10 and 2 columns of 10, thus removing 50 in all. But $100-50$ gives 50, which falls short of 7×8 by 6. We must find some excuse for adding 6, and it is close at hand. In the bottom right corner, we have 6 objects whose removal has been counted twice. As on page 243 we can tell a more realistic story; begin with 10^2; remove 2 columns of 10; add 6 objects at the bottom right corner; remove 3 rows of 10. We thus demonstrate the result:

$$(10-3).(10-2)=10^2-5.10+6$$

Compare this with:

$$(10+3).(10+2)=10^2+5.10+6$$

It will be seen that the same numbers, 5 and 6, occur in both results.

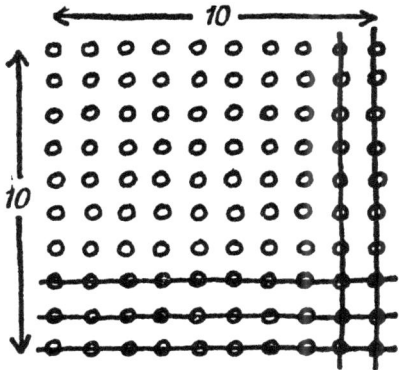

Figure 133

Simple apparatus to illustrate this connexion may be made in the following way. Circles are marked on rectangular pieces of card. If these circles are plain, they represent objects or *things*. If they are coloured red (as shown by shading in our illustrations) they represent *un-things*. An un-thing is, so to speak, a debt of one thing. If a thing meets an un-thing, there is a bang, and nothing remains. This idea actually plays a role in modern physics. A transistor may contain 'electron holes', which represent a shortage of an electron. Even more dramatic, there is the idea of anti-matter, which wipes ordinary matter out when the two meet. In Figure 134 A we can discern the various components of $10^2+5.10+6$. In Figure 134 B the 5.10 things have changed to 5.10 un-things, to represent -5.10. So B shows $10^2-5.10+6$. We now suppose the two flaps of un-things in Figure 134 B to

247

be turned inwards, so that they cover part of the 10 by 10 square (see Figure 135).

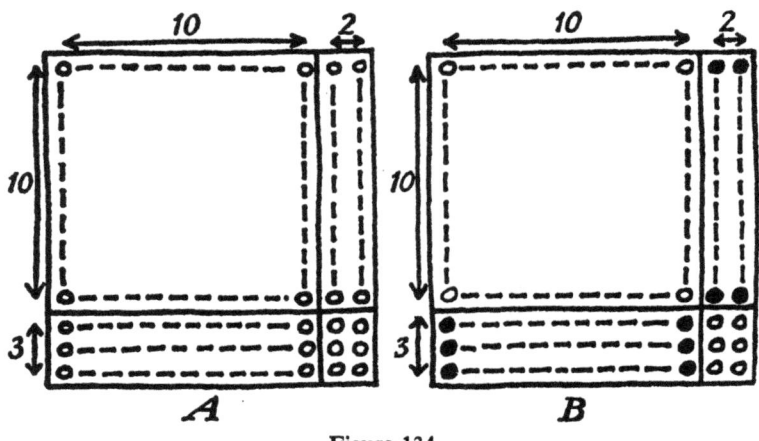

Figure 134

The card representing 6 things is then put on top of the region where the two flaps overlap. In the singly shaded regions, the things are covered by a layer of un-things, which cancels them

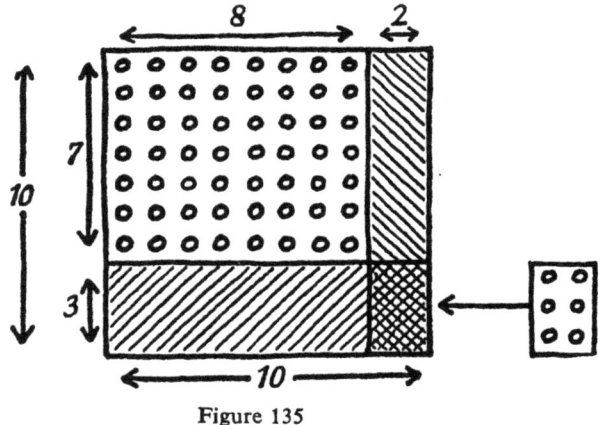

Figure 135

out. In the bottom right corner, we have first 6 things from the original square 10 by 10; then two layers of 6 un-things; and finally the 6 things on the small card. There are two layers of things and two layers of un-things; here again, the total is nothing. So the pieces shown in Figure 134 B combine to make exactly 7 rows of 8 things.

We arrive at Figure 134 B in illustrating $10-3$ times $10-2$. The various regions of 134 B illustrate the following numbers:

$$+100 \qquad -20$$
$$-30 \qquad +6$$

Here plus signs are put with 'things' and minus signs with 'unthings'. It will be seen that these numbers could be related to $10-3$ and $10-2$ by the following scheme in Figure 136.

Figure 136

We have arrived at this scheme with the help of pictures and lengthy considerations. If we are to perform calculations rapidly, we must once again make the transition to formal work. So we encourage children to observe the results above, and find shortcuts. Suppose we were given the numbers outside the box in the scheme above; how could we most quickly obtain the numbers inside? This question falls into two parts, which are best studied separately; (i) what numbers occur inside the box? (ii) what sign, $+$ or $-$, is attached to each number?

It is not hard to deal with (i). Each number inside the box comes from multiplying two numbers written outside. For example 6, inside the box, is obtained by multiplying 3 and 2; 3 is to the left, 2 is on top. We find 6 in the place where the row marked 3 crosses the column marked 2. The other numbers in the box are found in the same way. In fact, this part of the work is exactly the same as that done in multiplying $10+2$ by $10+3$. When you consider how closely Figure 134 B resembles Figure 134 A, this is not surprising.

Next we consider (ii), the question of the signs $+$ and $-$. For

249

this purpose, it is convenient to make a simplified version of the scheme above. We ignore the numbers, such as 10, 2, 100, 30, 20, 6, and note merely the signs + and −. The number 10, in the scheme above, is written without any sign at all. How shall we deal with this? We regard it as having a + sign. This 10 stands for 10 things, not for 10 un-things. It is a credit, not a debit. There is indeed a slight inconsistency in our usual way of writing additions and subtractions. If we write, say $100+7-13+2-9$, we might illustrate this by a story in which I began with £100, received £7, spent £13, received £2, and finally paid out £9. There are + signs for the amounts I received, − signs for what I paid out. No sign at all is written against 100, corresponding to the £100 I began with. But suppose I had begun £100 *in debt*. We would then have to write $-100+7-13+2-9$ to record my financial adventures. We cannot leave the minus sign out without falsifying the record. Now returning to the case where I begin the day with £100, it is clear that this £100 is an asset; it represents my balance from successful transactions on former days. So we might write it with a + sign, like this: $+100+7-13+2-9$. We would have to do this if we insisted that every number had either a plus or a minus sign with it, according to whether it represented money taken in or paid out. In everyday usage, a number with no sign specified has the meaning of a number with a plus sign. If I say 'I have £100' you understand that I have £100 *to my credit*; it would be most misleading to use this sentence to indicate that I was £100 in debt. If I say 'The temperature is 50°', you do not expect this to mean 'the temperature is 50° below zero'.

If we were told 'Someone has assets of £10 and £3 and owes £7,' we might be led to write $10+3-7$. Someone else might write $3+10-7$, which of course is equally correct. In the first form, the 10 is written without any sign but we see $+3$; in the second form the 3 receives no written sign, but $+10$ can be seen. If we wished to show explicitly which numbers represented assets and which represented debts, we could write $+10+3-7$ or $+3+10-7$. It is of course not usual to write it so. Both in mathematical work and in everyday arithmetic, it is usual to omit the plus sign in front of the number written first. *This can never be done with minus signs*; a debt of £2, a debt of £3 and a debt of £4 can be written −2−3−4, but we cannot suppress any of the minus signs without changing the meaning.

It is not surprising that some confusion arises, since plus and minus signs are used in two different senses in arithmetic. In kindergarten, we have 10−7 meaning 'begin with 10 and take 7 away'. But later we meet −7 by itself as meaning, say, 7° below zero. 'Take 7 away' is an order; '7° below zero' is a situation. In an expression such as (−7)−(−10), both meanings occur. The minus signs inside the brackets describe situations, 7° below zero and 10° below zero. The minus sign in the middle conveys an order; it says 'compare these situations; tell me how much hotter the first is than the second'. The answer is +3, and the plus sign here is not an order. It says the first situation is 3 degrees *hotter* than the second. And yet it is related to the command to add; for we could say that if you added 3° of heat to −10° you would get −7°.

To return to problem (ii), we can agree that 10−3 represents 10 things from which 3 are removed, crossed out, annihilated. The 10 is positive, the 3 negative. In Figure 137 is reproduced our earlier scheme; beside it we put the simplified scheme, in which numbers are ignored, and only signs are shown.

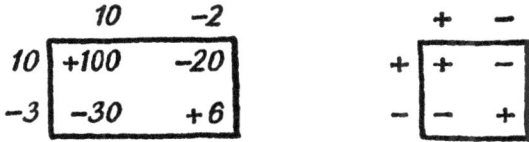

Figure 137

It will be convenient to call this last little table the multiplication table of + and −. It is not a large table. It has only 4 entries. Yet how much trouble it seems to cause! The difficulty probably does not lie in the table itself, but rather in the fact that rote-taught pupils have rules for adding, subtracting, and multiplying + and −, and these all get mixed together in one tremendous confusion. It pays to spend plenty of time discussing these basic processes, telling stories and drawing pictures, until each pupil can clearly visualize and distinguish the various operations.

In the little multiplication table above, the results 'plus times plus is plus' and 'plus times minus is minus' do not usually cause much surprise. The result, 'minus times minus is plus' often seems strange and unexpected. It is good if pupils think out the multiplication of $(10-2)$ by $(10-3)$ several times, with different stories or pictures.

We can work with mounds and hollows. Suppose that first of all a square has been made with 10 rows of 10 mounds. Then a message comes – there should be 2 less mounds in each row. We can cancel 2 mounds by providing 2 hollows (Figure 138).

Figure 138

If this is done in each row, the plan will be as in Figure 139.

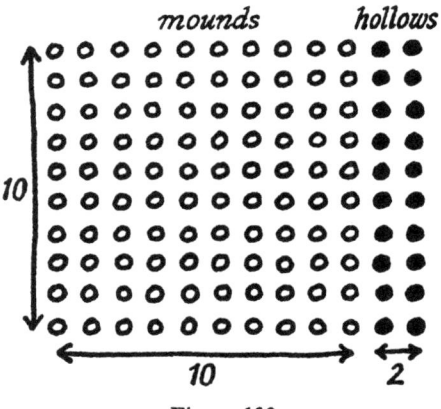

Figure 139

Now a further message comes – reduce the number of rows by 3. We want to cancel the last 3 rows. Each row contains 10 mounds and 2 hollows. To cancel 10 mounds we must provide 10 hollows, and to cancel 2 hollows we must provide 2 mounds. So we arrange for 3 more rows, each with 10 hollows and 2 mounds. The situation is now as in Figure 140, which is essentially the same as Figure 134 B.

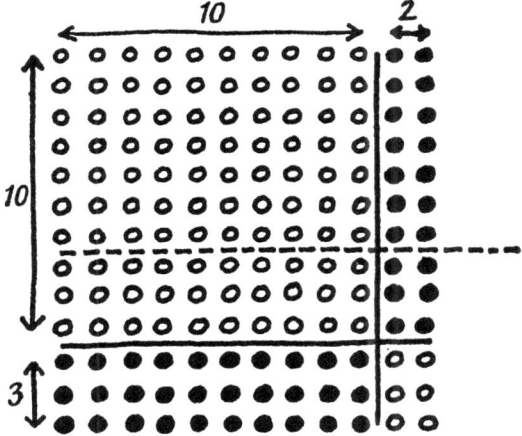

Figure 140

We can now imagine the cancellation carried out. First, everything below the dotted line can disappear – each row of 10 mounds being tipped into a row of 10 hollows below it, and the 6 mounds at the bottom right-hand corner being used to fill the 6 hollows above them. We cover the part of the drawing below the dotted line with a piece of paper. Now in each of the top 7 rows, we can use 2 mounds to fill 2 hollows. We are left with 7 rows of 8 mounds, illustrating $(10-3)$ times $(10-2)$.

The same multiplication can be illustrated in a commercial setting. A firm supplies 10 machines at a price of £10 each, and receives £100 in payment. In Figure 140 this £100 is shown by the square with 10 rows of 10. Money coming in is shown by unshaded circles, money paid out by shaded ones. However it appears that 3 of the machines are defective, so 3 refunds of £10 are sent out (the 3 shaded rows of 10, bottom left). It is then discovered that another department, because of some special

253

bargain offer, has sent out a refund of £2 to all 10 purchasers (the 2 columns of 10 shaded circles, top right). But the 3 customers whose machines were defective have had all their money returned. They are not entitled to the £2 refunds. So each of them has to send £2 to the firm (the 6 unshaded circles, bottom right).

The minus signs correspond to 'wiping out'. The defects wipe out 3 sales, so that $10-3$ machines are sold instead of 10. The refund of £2 wipes out part of the price; the machines are sold for $10-2$ pounds each. But when the 3 defective machines are returned, this not only wipes out the £10 payments, it also wipes out the £2 refunds. The wiping out of a refund has the effect of bringing money in. This is why $+6$ appears as part of the final answer when $10-3$ times $10-2$ is worked out in the way we have done.

Of course, this particular procedure for multiplication has little interest in arithmetic. It becomes useful when we pass from arithmetic to algebra, and instead of considering $10-3$ times $10-2$ consider $n-3$ times $n-2$. It is easy to replace 10 by n in Figure 134 b. The scheme for multiplication then becomes as in Figure 141.

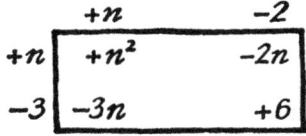

Figure 141

We thus have:

$$(n-2)(n-3)=n^2-5n+6$$

since $-2n$ and $-3n$ may be combined as $-5n$. The scheme above corresponds to Figure 134 b, with n instead of 10. The $+$ signs. correspond to 'things', the $-$ signs to 'un-things'. Or, if you like, the $+$ signs correspond to mounds, the $-$ signs to hollows.

We could perhaps find some way of adapting Figure 133 to illustrate this result. That is, we could draw circles, and cross them out when subtractions were made. However the mounds and hollows, or things and un-things, approach has certain

254

advantages. We may need to illustrate $5-2+3-4$. On the crossing-out approach we have to draw 5 circles, cross out 2, put 3 in, and cross out 4. The diagram becomes such a mess that it is impossible to see what is happening. But we can illustrate $5-2+3-4$ quite easily by either of the methods shown in Figure 142.

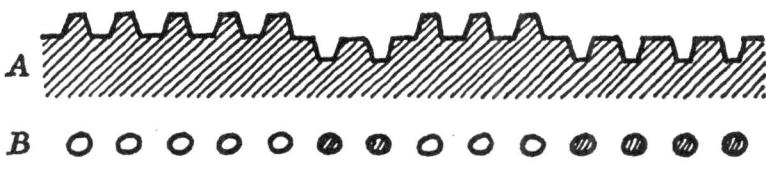

Figure 142

In Figure 142 B, you can think of the plain circles and the shaded circles as representing mounds and hollows, things and un-things, credits and debits, whichever you prefer.

If $5-2+3-4$ were being multiplied by $6-1+2$ we would begin by drawing 6 rows, each like Figure 142 B. When we reach -1 in $6-1+2$, we want to blot out one of our 6 rows. We could cross it out, but we prefer to provide a fresh row designed to cancel one of the rows already drawn. This fresh row would have to contain a mound wherever the earlier rows contained a hollow, and a hollow wherever the earlier rows contained a mound. Finally, for the $+2$ in $6-1+2$, we would have to draw two rows, like our original one. In this way, without erasures, we could show the whole history of the operation.

A diagram, an illustration of an idea, should not simply be something given by a teacher. The pupils should work the idea into their minds by themselves making diagrams and interpreting them. Painfully accurate draughtsmanship should not be required. Just enough should be drawn to make it clear that the idea is understood. Perhaps counters could be provided – white to represent mounds, things, and credits, red for hollows, un-things, and debts (in the red).

An array of such counters is shown in Figure 143. What does this array represent?

In the top row we see 5 mounds, then 2 hollows, finally 3

mounds. This row represents $5-2+3$. We see this row 4 times. Then there are 2 rows which represent a going into reverse – a wiping out of 2 of the earlier rows. Then we advance again; the bottom row gives us one more row like the top one. Thus we draw 4 rows, blot out 2, then add 1. The symbol for this is $4-2+1$.

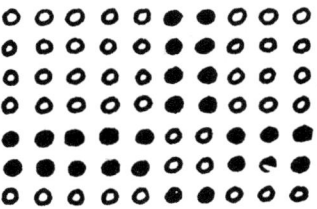

Figure 143

Thus Figure 143 represents the multiplication of $5-2+3$ by $4-2+1$. The figure breaks into rectangles. The number in each rectangle is shown in the scheme in Figure 144. Mounds are indicated by plus signs, hollows by minus signs.

	+5	*−2*	*+3*
+4	*+20*	*−8*	*+12*
−2	*−10*	*+4*	*−6*
+1	*+5*	*−2*	*+3*

Figure 144

The array contains in all 44 white counters (credits) and 26 red counters (debts). On balance, it represents $44-26=18$. Each row represents $5-2+3$, which is 6. As $4-2+1=3$, the total array should represent 3 times 6. As this is 18, we have a check on our accuracy and a confirmation of our way of thinking.

Children should use this type of check whenever they invent or interpret such an array.

As already mentioned, this procedure is a very elaborate and inefficient method of showing that $3 \times 6 = 18$. Its usefulness appears only when we come to algebra and have to multiply, say, $a-b+c$ by $p-q+r$. We merely need to reflect that there is nothing special about 5, 2, and 3; these numbers could be replaced by any numbers a, b, and c. In the same way, 4, 2, 1 may be replaced by p, q, r.

Accordingly we may arrive at the scheme for multiplying $a-b+c$ by $p-q+r$ (Figure 145).

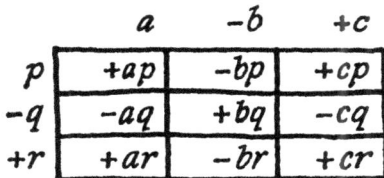

	a	$-b$	$+c$
p	$+ap$	$-bp$	$+cp$
$-q$	$-aq$	$+bq$	$-cq$
$+r$	$+ar$	$-br$	$+cr$

Figure 145

The balance of all the things written inside the box here gives the 'answer' for the multiplication. It is a long and wearisome answer, and we may well wonder how much it will help us in solving problems. But at any rate, for whatever it is worth, it is the correct answer.

The essential point is that we are now able to deal with both plus and minus signs. The method of multiplication described in Chapter 9 applied only to sums; all the signs had to be plus. Now we are free from that limitation.

Suppose for example we wish to multiply x^2-2x+3 by x^2-4x+5. We draw our scheme (Figure 146).

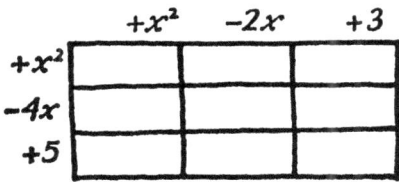

	$+x^2$	$-2x$	$+3$
$+x^2$			
$-4x$			
$+5$			

Figure 146

Then we have to ask ourselves (i) how many counters in each rectangle, (ii) are they white $(+)$ or red $(-)$? To answer question (i), we need only the ideas already discussed in Chapter 9. We have to know that x^2 times x^2 is x^4, that x^2 times $2x$ is $2x^3$, and so forth. To answer question (ii) we have to understand and remember 'the multiplication table of plus and minus'. Writing in each rectangle the number of counters, with the appropriate sign, we obtain the scheme in Figure 147.

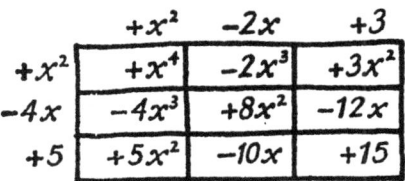

Figure 147

Finally, we can write the expression inside the box in a shorter way. Just as in Chapter 9, there is no point in having $5x^2+8x^2+3x^2$. This can be shortened to $16x^2$. In the same way, $-2x^3-4x^3$ may be written more shortly as $-6x^3$. At this stage of the work, we are performing addition. We know that the answer to the multiplication is found by *putting together* the 9 things written inside the box above. So we may expect children to make all the mistakes mentioned in Chapter 9 under the heading of addition. They may want to lump together x^4 and $-2x^3$ and get $-x^7$ or some other incorrect result. If they successfully avoid all these errors, they should arrive at the answer $x^4-6x^3+16x^2-22x+15$.

In fact to carry out a multiplication such as that just done, the pupils must understand the following four procedures:

   (i) How to set out the rectangle scheme for multiplication;
   (ii) How to carry out the multiplication of single terms, such as $5x^3$ times $2x^4$;
   (iii) How to get the correct sign, plus or minus, written in each rectangle;
   (iv) How to add together all the different parts of the answer, without falling into errors.

It will be seen that children have to understand and remember four different ideas, and use these correctly in one large-scale operation. This calls for a considerable degree of mental organization. It is not to be expected that all children will rapidly or easily learn to do this. In teaching we need to be patient. We have

The Routines of Algebra II

to be careful not to present so many ideas at once that children fail to grasp any of them. All the time we have to probe and check what each child understands. If a question is wrongly answered, we try to discover whether this is due to a mere slip, or whether it comes from a misunderstanding of one of the four items just listed. Naturally, we question the child on each principle separately, and see just where the weakness lies.

Exercises and Examples

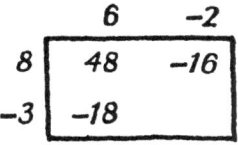

Figure 148

1. The scheme in Figure 148, for multiplying $6-2$ by $8-3$, needs to be completed. As $6-2=4$ and $8-3=5$, the numbers inside the box should combine to make 5 times 4, that is, 20. Do we need -6 or $+6$ in the remaining space for this to happen?

2. What is the missing number in the scheme in Figure 149? What numbers are being multiplied? Is the answer correct?

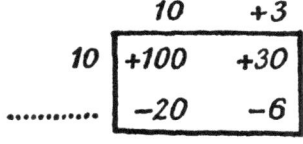

Figure 149

3. In this scheme (Figure 150), $x+3$ is being multiplied by $x-2$. How would the final answer be written? How is this question related to question (2)?

$$
\begin{array}{c|cc}
 & x & +3 \\
\hline
x & +x^2 & +3x \\
-2 & -2x & -6 \\
\end{array}
$$

Figure 150

259

4. What multiplication is being done here (Figure 151)? How would the final answer be written?

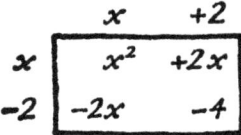

Figure 151

5. Complete the scheme in Figure 152 and give the answer for $(x-3).(x+3)$ in its simplest form.

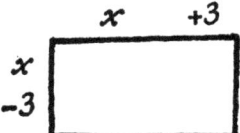

Figure 152

6. Questions (4) and (5) dealt with $(x-2)(x+2)$ and $(x-3)(x+3)$. The results suggest that it might be worth while to multiply out $(x-4)(x+4)$ and $(x-5)(x+5)$ and so on. Do this. What do your results suggest $(x-a)(x+a)$ should be, where a stands for any number?

7. Complete the scheme in Figure 153 for multiplying $x+a$ by $x-a$, and give the answer in its simplest form. Does your answer confirm the guess you made in question (6)?

Figure 153

8. Multiply $x+1$ by $x-1$. To check your answer, take the particular number 10 for x, 'the number thought of'.

9. Multiply x^2+x+1 by $x-1$. Again check with 10 replacing x.

10. Multiply x^3+x^2+x+1 by $x-1$ and check as in the previous questions by using 10.

11. We know that $11,111 \times 9 = 99,999 = 100,000-1$. In view of what we have done in questions (8), (9), (10), this suggests a certain multiplication in algebra, and its answer. What is the multiplication, and is the answer suggested correct?

12. Do you think there would be any multiplication question in algebra, in which $x-1$ times something gave the answer x^6-1? What would the 'something' have to be?

13. Complete the scheme and find the answer for $(x^2+x+1)(x^2+x+1)$ (Figure 154).

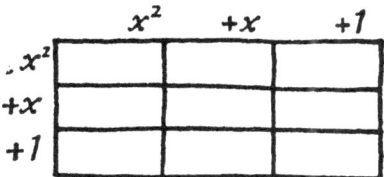

Figure 154

14. Complete the scheme in Figure 155 and find the answer for (x^2-x+1) (x^2-x+1). (Compare the pattern made by the signs $+$ and $-$ with the schemes on pages 256–8.)

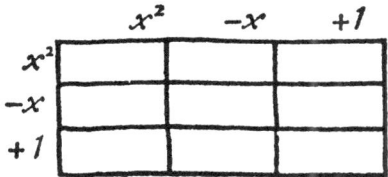

Figure 155

15. Complete the scheme and find the answer for (x^2-x+1) $\times(x^2+x+1)$. Check your answer by replacing x by 2. (It is possible to set out an array of white and red counters to illustrate $4-2+1$ times $4+2+1$. If you replace x by 2 in your completed scheme, the resulting numbers should correspond to rectangles of counters in the array.)

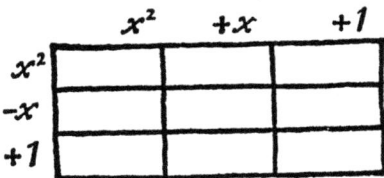

Figure 156

16. Work out the following three multiplications:

(i) x^2-y^2 times x^2-y^2

(ii) $2xy$ times $2xy$

(iii) x^2+y^2 times x^2+y^2

Check that adding together your answers to (i) and (ii) gives your answer to (iii).

If any numbers are chosen for x and y, what would you expect to be the shape of a triangle with sides x^2-y^2; $2xy$; x^2+y^2? (We suppose the number chosen for x is larger than that chosen for y, so that x^2-y^2 is not 'in the red'.)

Could you choose x and y so that the sides of the triangle would be 3, 4, 5?

DIVISION

Anyone who understands multiplication understands division. If in arithmetic we are asked 'How many sevens in 42,' this is a problem in division – but we do not have to learn division tables. We run through the 7 times table in our mind, until we hit on '6 sevens are 42'. Then we have our answer. Many people, of course, do not know that 6 sevens are 42, and for them this division question will seem almost impossible. There are ways of getting round the difficulty. The method explained on page 213 might perhaps help them. However, as a general rule, a person should not be expected to do questions in division that lie beyond his knowledge of multiplication. A person should not be asked to divide by 7 if he does not know how to multiply by 7. For we test the correctness of a division by multiplying. If we want to

The Routines of Algebra II

check the result $112 \div 7 = 16$, we multiply 16 by 7. If we cannot do this multiplication with confidence, we are in danger of committing the mortal sin of mathematics – making statements which we cannot test for ourselves.

It should be emphasized that division is a kind of guessing. How many sevens in 42? We can only try results in the 7 times table until we find one that works. Could it be 10? No, 10 sevens are 70; too big. What about 5? No, 5 sevens are 35; too small – but not much too small. Creep up a little – 6 sevens, yes, that is it. In any procedure where guessing is involved the skill lies in *testing the guesses*. A first guess may be wrong; that is no discredit. In science, a first guess nearly always is wrong. A good scientist is someone with the courage to guess wrong, and then see in what direction the guess ought to be amended. The absolutely essential thing is a satisfactory way of testing the guesses. Otherwise we may have a wrong answer, and rest content with it.

Since division is tested by multiplication, we have a choice of two procedures in teaching division. One way would be not to teach any division at all until the pupil was expert in every type of multiplication. The other way is to teach division and multiplication together, but to see that the division questions are always based on types of multiplication which the pupil has thoroughly mastered.

This second method is more satisfactory, and applies at all levels of teaching. When a child knows that 3 twos are 6, we may ask 'How many twos in 6?'. If a child has done enough multiplication of the type $(x+2)(x+3) = x^2 + 5x + 6$, it would be reasonable to ask him what $x+3$ must be multiplied by to give $x^2 + 5x + 6$.

Now of course division is not always exact. Our question may be: 'How many sevens in 45?' and the answer: that there are 6 sevens, and 3 to spare. In effect, the question is, 'How close can you get to 45 in the 7 times table?' But here we meet a difference between arithmetic and algebra. In arithmetic, we get as close as we can, but always staying below the number. For example, if we were asked to divide 599 by 100, we would say that there are 5 hundreds in 599, and 99 left over. But of course we could get closer to 599 if we took 600. We could say that 599 contained

263

6 hundreds, all but 1 unit. This would be like saying that 100 goes into 599 six times, with a remainder of -1. This would be a most unusual answer in arithmetic, since arithmetic is closely related to our experience of physical objects. We cannot say 'I have 599 things. I will give each of my 6 friends 100 things, and that leaves me with 1 un-thing.' If we lived in some part of the universe where matter and anti-matter abounded, we might have become accustomed to doing this.

In algebra, the distinction between positive and negative carries much less weight than in arithmetic. If we represent $+3$ by 3 mounds and -2 by 2 hollows, we do not feel that one picture is much more complicated than the other. A mound and a hollow are equally familiar ideas. So in algebra we are not greatly disturbed if a minus sign occurs in an answer.

This point can arise in a very simple problem such as 'Divide $3x+2$ by $x+1$'. This problem can be visualized by the bags-and-stones method of Chapter 4. How many times does a bag and a stone go into 3 bags and 2 stones? From the viewpoint of arithmetic we would be inclined to answer 'Twice'. Starting with 3 bags and 2 stones, we could give each of 2 people a bag and a stone. Here we should have to stop, for although a bag is left on our hands, we have no stone to put with it. Our remainder would thus be a bag, or x in more conventional language.

On the other hand, we can very nearly hand out a bag and a stone to each of 3 people. That would leave us in debt to the extent of 1 stone. So we are led to consider two possible answers to the question 'Divide $3x+2$ by $x+1$'.

Possibility (i): It goes 2 times with remainder x.

Possibility (ii): It goes 3 times with remainder -1.

Which answer is correct? It depends on circumstances. If this division arose when you were solving a problem about physical objects, where you knew that it would be ridiculous to have minus numbers, then possibility (i) might be appropriate. This, however, would be rather exceptional. In algebra as a rule, possibility (ii) would be what we wanted. If the question was asked by a school teacher or came in an examination paper, without reference to any particular problem, you could be certain that '3, with remainder -1' was the answer desired.

The Routines of Algebra II

In division, we want the remainder to be as simple as possible. We could answer the question 'How many sevens in 100?' by saying '3, and a remainder of 79'. In a strange kind of way, this answer checks, since $100 = 3 \times 7 + 79$. But we regard '14, with remainder 2' as the proper answer since a remainder of 2 is so much simpler than a remainder of 79. But it is not so easy to decide between a remainder of x and a remainder of -1. Which is it better to have, a negative number, -1, or an unknown number, x?

Really, the issue here is that the person asking the question should make clear what he wants. In arithmetic, we are sometimes uncertain what a question means. If you ask a large number of children 'What is 5 divided by 2?' some will answer '$2\frac{1}{2}$'; others will answer '2 and 1 over'. Both are reasonable answers. The answer $2\frac{1}{2}$ would be appropriate to a question about 2 children sharing 5 cakes, but highly inappropriate to a question about dividing 5 players into teams for some game played with 2 sides. When such a question is asked purely as an exercise in arithmetic, the person posing the question should be very careful to indicate which type of answer he requires. Often this is not done. In algebra there is even more scope for misunderstanding, and it is the responsibility of the questioner or examiner to indicate what sort of expression he regards as permissible for an answer.

There is a widely accepted convention as to what constitutes the 'simplest' remainder for division in algebra. We regard x as being more complicated than any fixed number, positive or negative, whole or fractional. Thus we would prefer any of the remainders -1; 10; $-7\frac{1}{4}$; or 2,073,492$\frac{1}{2}$ to x. We regard x^2 as even worse than x. We would sooner have a remainder of 1,000 x than one of x^2. In the same way, something involving x^3 would be worse than anything with x^2.

It will be seen that this agrees with our earlier ruling, that '$x+1$ goes into $3x+2$ three times, with remainder -1' is the generally accepted statement.

Our dislike of the higher powers of x determines our strategy in division. Rather as in Chapter 9, we form rows, always trying to fit the highest power of x. We stop when we cannot make 'what is left' any simpler. The difference now is that we consider problems in which minus signs are involved. These minus signs may not always appear in the question. For example, we may wish to divide x^2+x+1 by $x+2$. Considering region A (as on page 216) 'we see that 'x rows of $x+2$' would be a good start. Now x rows of $x+2$ contain x^2+2x objects, and we have gone into debt. If we keep our accounts as on page 215 we have;

Initial amount	x^2+x+1
x rows of $x+2$ objects use up	x^2+2x
and leave	$-\ x+1$

The remainder, $-x+1$, still contains x, so we should like to continue and get rid of this if possible. Because of the minus sign, 'un-rows' are now in order. In fact, if we take simply -1 row of $x+2$ objects, that will do the trick. For -1 times $x+2$ is $-x-2$, and that makes the x term what we want.

If we went on with our accounts in the same way, the next step would be to subtract $-x-2$ from $-x+1$, the amount left over from the first step. This is the usual routine, and it is perfectly satisfactory for many people. Others find they get confused, and lose sight of what is happening. For anyone in this position, the wisest thing is to start the argument again. We want to find how many rows of $x+2$ objects will get us closest to x^2+x+1. We found that x rows gave too much, and that it seemed wiser to consider $x-1$ rows. Let us see how close this has got us to x^2+x+1. By the methods already explained, we find that $x-1$ rows of $x+2$ objects makes x^2+x-2 in all. This is 3 less than x^2+x+1, so from x^2+x+1 objects it is possible to make $x-1$ rows of $x+2$, and have 3 left over. This settles our division question; the answer is $x-1$ with remainder 3.

It will be seen that the essential processes in division are multiplication and subtraction. We guess how many rows are needed. We multiply to see how many objects those rows contain. We then subtract to see how far we are from our goal. We use the result of this subtraction to make another, better guess. We continue in this way, until we are satisfied that we have the best possible guess.

There is much to be said for pupils approaching division in this way. The usual routine procedure simply allows a certain saving of writing, and can be regarded as a shortened version. Children write so much nonsense in long division that one would prefer the slightly longer method, which is certain and is based on understanding, to the shorter method, which is often mangled into absurdity. It may well be possible to teach the guessing method first, and later show how it may be shortened and stream-lined for professional use.

The work below shows the successive guesses for a typical question involving minus signs.

Question: divide $3x^3-10x^2+13x-29$ by $x-2$. The terms in heavy print suggest our first guess, that we want $3x^2$ rows. As $3x^2$ times $x-2$ is $3x^3-6x^2$ we are in error by $-4x^2+13x-29$. The first term (in heavy print) suggests that we amend our guess, by bringing in $-4x$ rows.

Second guess: $3x^2-4x$ rows of $x-2$ objects. We are led to the multiplication in Figure 157.

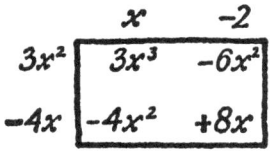

Figure 157

The total inside the box is $3x^3-10x^2+8x$, so we are still in error by $5x-29$. This suggests that we try 5 more rows.

Third guess: $3x^2-4x+5$ rows of $x-2$. Our multiplication now is shown in Figure 158.

	x	-2
$3x^2$	$3x^3$	$-6x^2$
$-4x$	$-4x^2$	$+8x$
$+5$	$+5x$	-10

Figure 158

The total in the box is now $3x^3-10x^2+13x-10$. We were hoping to reach $3x^3-10x^2+13x-29$. The difference is -19, and we cannot get any simpler remainder. So we have found that $3x^3-10x^2+13x-29$ is $3x^2-4x+5$ times $x-2$, together with -19. So $x-2$ goes $3x^2-4x+5$ times and the remainder is -19. Pupils may well be in doubt whether the remainder is -19 or 19. They see that the two expressions differ by 19 but may not be sure which should be taken from which. It is for this reason that we put the full statement above – so many rows of $x-2$, together with -19. The remainder is found after the words 'together with'. Compare the statement for 45 divided by 7; 45 is 6 rows of 7 *together with* 3.

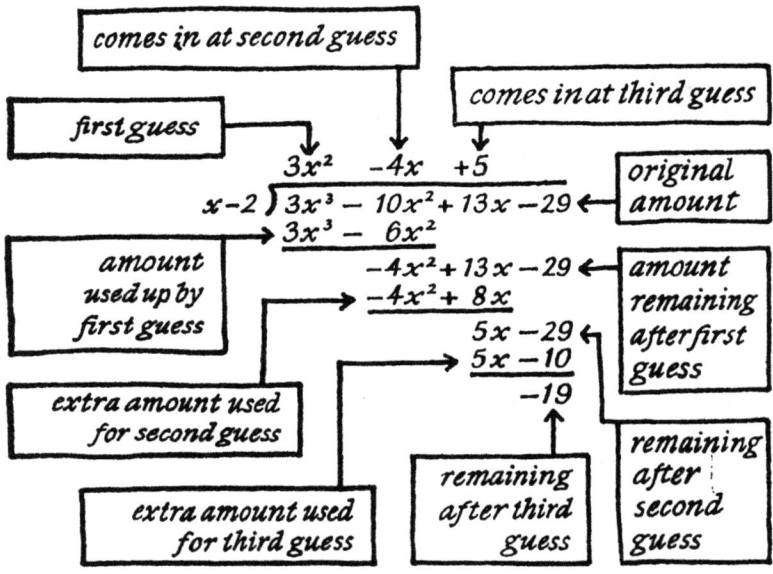

Figure 159

The full-dress argument just given is the safest for anyone whose algebra is weak. It clearly can be shortened for practical use in more expert hands. First, we notice that it is unnecessary to go back to the beginning each time we multiply. The last box shown above has 3 lines. The first 2 lines are exactly the same as the box we had at the stage before. So here is a very simple and obvious economy. We use only one box and allow it to grow by a line as each new guess is made. The second economy is concerned with finding the 'errors' or remainders. At each stage of the guessing, a fresh line is added to the box. The material for this line comes from the remainder left by the previous stage. Accordingly, the new remainder can be found by subtracting from the old remainder what has gone into the extra line. This is in fact the exact procedure used in the accounting on page 218. We have in fact returned to the method used in the simplest questions, where no minus signs appeared. The only reason for the extra discussion is that children (and others) are very liable to become confused when the using up of minus amounts is concerned.

Figure 159 shows the usual way in which the division performed above would appear in a textbook. It also gives labels to indicate what each part corresponds to in the guessing procedure.

However, there is no urgency for getting children to use this particular form. It is far better for them to use any procedure which they have thought out for themselves, and which they can use correctly, than to attempt some streamlined procedure which they apply in a hit-or-miss fashion with a large proportion of incorrect answers.

CHAPTER ELEVEN

Graphs

GRAPHS rest on one of the simplest and most useful ideas in mathematics. A graph will often help us to solve a problem, and children should learn to regard graphs as a useful aid, and to sketch a graph whenever occasion arises. It is a great pity when graphs are taught ponderously and unattractively, as a subject in themselves, so that pupils come to dislike and avoid graphical work.

The introduction to graphical work can be quite informal. A class of young children are confronted with the following problem. There are two aeroplanes, a faster one that flies at 10 miles a minute and a slower one that flies at 5 miles a minute. The slower plane has a start of 20 miles, and the faster one is chasing it. When will the faster plane overtake the slower one?

It is quite natural to start drawing pictures (see Figure 160).

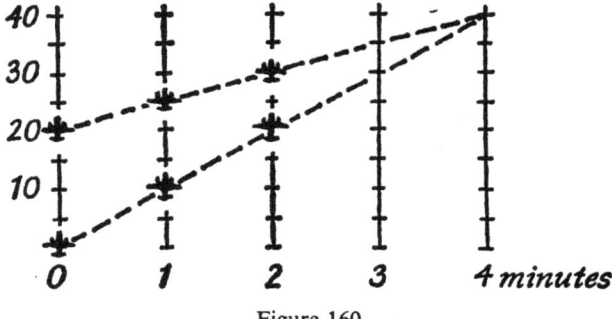

Figure 160

The first line shows the situation at the beginning of the chase. The other lines show the positions of the planes after 1, 2, 3, 4 minutes. In Figure 160 the positions of the planes have been shown only for the first three times, corresponding to 0, 1, and 2 minutes. Already children will notice that the marks lie in line (as shown by the dotted lines) and they will predict that over-taking occurs after 4 minutes, corresponding to the point on the graph where the dotted lines cross. It is not even necessary to use graph paper. A rough sketch on paper or a blackboard may well be sufficient to suggest the dotted lines, if the drawing is reasonably accurate. Squared paper may be used, but if so one should not make a meal of it. Minute attention to the mechanics of plotting points is definitely out of order. The aim is to picture the problem quickly, and to predict when the chase will end. Once we have guessed that 4 minutes is the required time, it is easy to check that this is correct. For in 4 minutes the faster plane will cover 40 miles; the slower one will cover 20 miles, and as it had a start of 20 miles, will also be at the 40 mile mark. The graph thus appears as an aid to guessing, which is essentially what it is.

The children may make up similar problems for themselves, and draw the graphs. Without being told by anyone, they will soon come to realize the useful fact that the graph of an object moving at a fixed speed is always a straight line.

Young children delight in codes and secret messages. This is another direction from which graphs can be approached. The code is a very simple one, as in Figure 161.

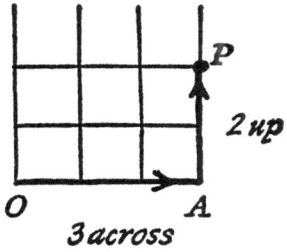

Figure 161

Suppose I wish to convey to someone that I am thinking of the point *P* in Figure 161. If I begin at *O*, to reach *P* I have to travel 3 steps across the paper, and 2 up. We always begin at *O*, the south-west corner of the paper. The two numbers, 3 and 2, completely fix the position of *P*. We agree that the number *across* is

271

always given first, and the number *up* second. This has to be remembered, and of course a little practice is needed to fix it in the memory. Our code name for the point *P* is thus (3,2).

It may need to be emphasized that all code names for points are based on our starting at the corner, *O*. Sometimes, when two or more points are named in code, children tend to measure on from the point last named, or they may be in doubt whether this is what we intend them to do. However, that is not our intention. For instance in Figure 162, after marking (3,2) at *P*, children may

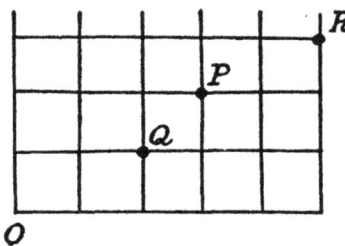

Figure 162

want to mark (2,1) at *R*, since *R* is 2 across and 1 up from *P*. But we always work from *O*. The point (2,1) is not *R* but *Q*. The code name of *R* is (5,3).

Children can have a lot of fun with these codes. A child can draw

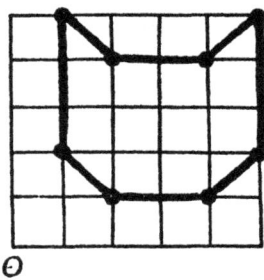

Figure 163

a picture, turn it into code, then other children can try to turn the code back into the picture.

For example, Figure 163 is intended to represent a cat's face.

272

The code instructions might read, 'Join (2,1) to (4,1) to (5,2) to (5,5) to (4,4) to (2,4) to (1,5) to (1,2) to (2,1).' If a teacher has followed these instructions and drawn the cat's head on the blackboard a discussion might follow as to where the cat's eyes ought to be. Children do not need to come to the board. They can call out their suggestions in code, for example, 'The eyes should be at (2,3) and (4,3)'.

This type of work does not lead to any particular mathematical result, but it makes children familiar with the system of specifying points by numbers. The learned name for these numbers is 'the coordinates of the point', and coordinates are widely used in mathematics. The name need not be mentioned to children; 'coordinate' is a long word for a simple idea. But all kinds of questions can be discussed. If you were at (3,4) and moved across (to the east), what would be the code names of the points you passed through? If you started at (5,1) and moved up (north), what points would you pass through? If you began at O and moved north east, what points would you pass through? What do you notice about the numbers for them?

The code idea can also be used for actual messages. The dots on the graph paper can be made to form letters.

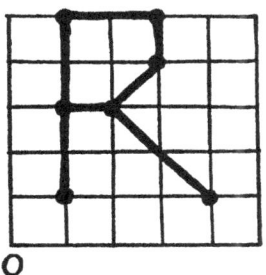

O

Figure 164

Figure 164 shows the letter *R*. The code instructions might read, 'Join (1,1) to (1,5) to (3,5) to (3,4) to (2,3) to (1,3). Also join (2,3) to (4,1).'

In the diagrams so far we have not used points on the edge of the squared paper. Sometimes we may want to mention such points. In Figure 165 if we start at *O*, to reach *A* we have to go 3 across

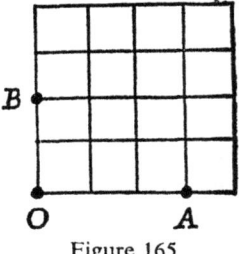

Figure 165

and 0 up. So *A* is written (3,0). To get to *B* from *O*, we go 0 across and 2 up, so *B* is written (0,2). If we want to mention *O* itself, we write (0,0), for to get from *O* to *O*, we go 0 across and 0 up.

THINGS TO NOTICE

In Figure 166 we have drawn a line and marked several points on it, dotted about in a rather random order.

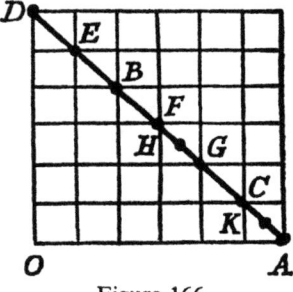

Figure 166

Consider the code names of a few of these points. The point *A* is (6,0); *B* is (2,4); *C* is (5,1); *D* is (0,6); *E* is (1,5); *F* is (3,3); *G* is (4,2). Do we notice anything about these numbers? Children usually notice fairly quickly that the two numbers inside the bracket add up to 6. (They may state this result in some equivalent form, for example, that you can get the second number by subtracting the first number from 6.)

In order to state this result briefly, we bring in abbreviations. Let x be short for 'the number we go across to get to a point', and y for 'the number we go up to get to the point'. That is, the code name of the point is (x,y); x is short for 'the first number in the bracket, y is short for the second number in the bracket'. But we have noticed that, for the points A, B, C . . . G on the line, these two numbers add up to 6. That is $x+y=6$, for each of these points.

This raises two questions. (i) If we take any other point on the line, shall we find for it too that $x+y=6$? (ii) Can we find any point not on the line that has $x+y=6$? Needless to say, we do not answer these questions. We ask the children to investigate. They might suggest that we try the point H, which is $(3\frac{1}{2},2\frac{1}{2})$, or K which is $(5\frac{1}{2},\frac{1}{2})$, or other points on the line. In each case, we find that $x+y=6$. With H, for example, $x=3\frac{1}{2}$ and $y=2\frac{1}{2}$, and these numbers do add up to 6. So far as we can judge by experiment, every point on the line has $x+y=6$. What, then, about the second question; can $x+y=6$ ever happen when the point (x,y) is not on the line? Again the children should suggest points, and try what happens. In Figure 167, 4 points, P, Q, R, S have

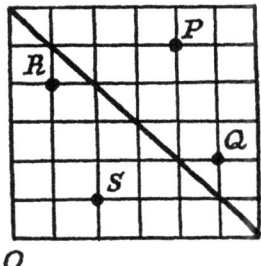

Figure 167

been chosen off the line. P is $(4,5)$ and for P the 2 numbers add up to 9. For Q, which is $(5,2)$ we get the sum 7. For R, which is $(1,4)$ we get the sum 5. For S, which is $(2,1)$ we get the sum 3. If children take many points they will probably notice that for points above the line, the sum $x+y$ is always more than 6, while for points below the line $x+y$ is always less than 6; only on the line is $x+y$ equal to 6. They may also notice that the nearer the point is to the line, the nearer $x+y$ is to 6.

So the statement $x+y=6$ is very closely connected with the line in Figure 166. It is true when the point (x,y) lies on that line, and only then is it true. So mathematicians use $x+y=6$ as a way of specifying that line. You can tell a child that if he went up to a mathematician in the street and said, 'I am thinking of the line $x+y=6$', the mathematician would know exactly what the child meant. He would know the child was thinking of the line shown in Figure 166.

Different children may express what they notice in different ways. Instead of saying 'The two numbers add up to 6,' they may say, 'If you take one number away from 6 you get the other.' This of course expresses the same observation in a slightly different way. In our picture language of Chapter 5, it does not matter whether you say that the boy standing on the stool is as tall as the man ($b+s=m$ for short), or that the boy's height subtracted from the man's gives the height of the stool ($m-b=s$ for short), or that the stool's height taken from the man's gives the boy's ($m-s=b$ for short). All three statements correspond to the same picture; they are different ways of saying the same thing. If a child says to the mathematician, 'I am thinking of the line $y=6-x$' or 'I am thinking of the line $x=6-y$', the message is exactly the same as that conveyed earlier.

We began with the line drawn on graph paper, and from it we arrived at the equation $x+y=6$. Sometimes it is necessary to go in the opposite direction. A mathematician may say to us, 'I am thinking of $x+y=4$'. What has he in mind? Suppose we take any point of the graph paper (see Figure 168), say A, which is (4,3). As $4+3=7$, the statement $x+y=4$ is not true for A.

We put a blob of yellow paint on A (shown in Figure 168 by a hollow circle). Yellow is for points that fail the test, black for those that pass. Next we try B, which is (3,1). Now 3 and 1 do add up to 4, so the statement $x+y=4$ is true for B. We put a blob of black paint on B. For C also, which is (2,2), we find $x+y=4$ to be true, so C is painted black. We try D, which is (2,3).

Graphs

As 2 and 3 do not add up to 4, *D* gets a yellow mark. But *E*, being (1,3), has two numbers which do add up to 4, so *E* is painted black. If we continue in this way, we eventually obtain a drawing in black on a yellow background. The black points give us the graph of $x+y=4$. These are the points for which the statement is true that the number across and the number up give 4 when added.

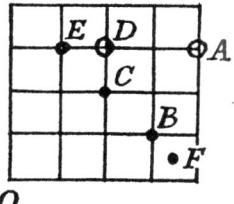

Figure 168

We need not restrict ourselves to whole numbers. A point such as *F*, with code name $(3\frac{1}{2},\frac{1}{2})$, is marked black.

You will notice that once again we have a straight line for our graph, and it has the same direction as the earlier line, from north west to south east. At this point, all sorts of questions should spring to children's minds. If we took some other straight line, running from north west to south east, would we get an equation '$x+y=$something' for it? What about lines in other directions? Does every equation have a straight line for its graph? The more children can be encouraged to propose and investigate such questions the better it will be for their mathematics, and the more interesting the work will be. Finding out is exciting; being told is dull.

We may experiment with a line in a different direction, for example the line shown in Figure 169.
We may tabulate as follows:

Point	A	B	C	D	E
Number across (x)	0	1	2	3	4
Number up (y)	1	2	3	4	5

What can we notice about these numbers? You may find that the

277

earlier work with $x+y=6$ and $x+y=4$ has put the children's minds in a groove. They become obsessed with *adding* x and y. They will say 'For A, $x+y$ is 1; for B it is 3; for C it is 5. Oh yes, we are getting the odd numbers.' Now the trouble, so to speak, is that this is perfectly true. We do get the odd numbers. But this is something which often happens in trying to solve a puzzle; we notice something, which is not quite what we want, and this blinds us to some much simpler observation, which in fact would

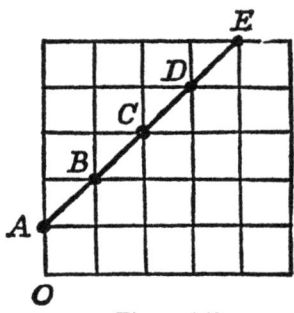

Figure 169

solve our puzzle for us. In this particular case, we may draw a dividend from the guessing game described in Chapter 4. Regard x as 'the number called out', and y as 'the number answered'. You call 1, I answer 2. You call 4, I answer 5; what am I doing? I am adding 1 to the number you say. The rule that fits the table above is in fact $y=x+1$. Further evidence can be obtained by considering fractions; for example $(\frac{1}{2}, 1\frac{1}{2})$, $(3\frac{1}{2}, 4\frac{1}{2})$, $(1\frac{1}{3}, 2\frac{1}{3})$ are 3 more points on this line.

A graph can be drawn to illustrate any of the guessing games described on page 79. There may be a slight practical difficulty with some of these; in games 4 and 5 for example rather large numbers occur, and it may be rather inconvenient to graph these. It is not impossible; such graphs can be made, if a large sheet of paper is used, and the children have enough patience and courage. The teacher should use his judgement. It is bad to do work which the pupils find so tedious that the intellectual excitement of the investigation becomes destroyed. On the other hand, children should not be given the impression that something tedious is impossible. We could most certainly draw the graph of game number 5 if it became necessary in some industrial or scientific problem, where we were determined to obtain a solution. In a class of children, there may be some enthusiastic or obstinate pupils, who find some degree of difficulty a challenge. They perhaps might draw the more tedious graphs, and show the results to the rest of the class. This procedure corresponds to the needs of modern society. We cannot all be experts on everything. An efficient society has many specialists, but encourages these specialists to work harmoniously together, and to keep the whole community informed of what they are learning.

All the graphs we have mentioned in this chapter so far have been straight lines, and most of the graphs that arise from guessing games are likely to be straight lines. There is a danger that children will jump to the (false) conclusion, that all graphs are straight lines. They may then fall into the habit of getting two points on the graph, then putting a ruler against these and drawing a straight line. It is therefore essential, fairly early in the teaching of graphs, to raise the question, 'Does every equation lead to a straight line?' and to give at least one example of an equation that does not. A simple example would be the equation $xy = 12$. Here the test is: do the two numbers used in writing the point multiply together to give 12? Thus (2,6) would pass the test, since $2 \times 6 = 12$, but (3,5) would fail and get a yellow blob, since 3×5 is not 12. It is easily seen that the following points all pass the test; (1,12), (2,6), (3,4), (4,3), (6,2), (12,1). It is quite clear from Figure 170 that these points do not lie on a straight line. There are of course other points on the graph besides those shown in Figure 170, for example $(8,1\frac{1}{2})$ and $(1\frac{1}{2},8)$, not to mention $(48,\frac{1}{4})$ and (0·1,120).

If we plot many points, they will begin to appear like the continuous curve shown in Figure 171.

CIRCLES REGARDED AS GRAPHS

A simple curve can have a fairly complicated equation. A circle, for example, is a very well-known curve, but we have to write a fairly long equation to specify a circle on graph paper. Such equations can be found by a quite straightforward procedure, which we do not want to discuss at the moment, as it would distract pupils from the basic ideas of graphs.*

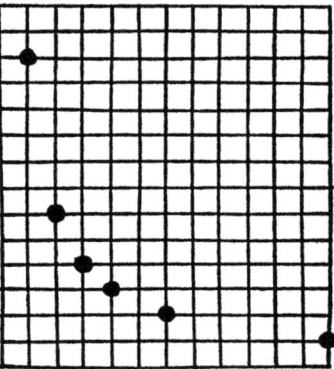

Figure 170

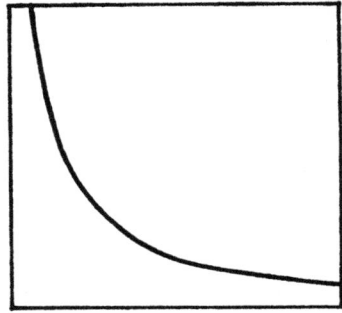

Figure 171

For example, the equation $x^2+y^2+25=10x+10y$ represents a circle. It is not too easy to guess numbers for x, y that 'pass the

* The equation of the circle with centre (a,b) and radius r is in fact:
$$x^2+y^2+a^2+b^2=2ax+2by+r^2$$
If we replace a, b, and r by particular numbers we can obtain as many examples of the equations of circles as we wish.

test'. Pupils, however, can check for themselves the statement that all the points listed here do qualify, and should be marked in black: (5,0); (0,5); (5,10); (10,5); (2,1); (1,2); (8,1); (1,8); (9,2); (2,9); (8,9); (9,8). For example, to test that (9,8) passes the test, we replace x by 9 and y by 8 in the equation.

Then $x^2 + y^2 + 25 = 10x + 10y$
becomes $81 + 64 + 25 = 90 + 80$

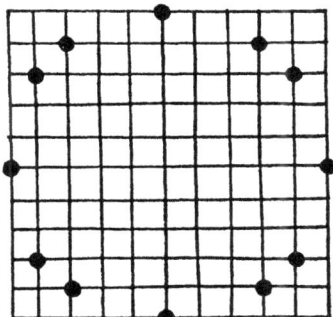

Figure 172

which is a true statement, since it says $170 = 170$. The other points may be tested in the same way. These points are shown in Figure 172 and it will be seen that they do suggest the shape of a circle. There are of course many other points that pass the test, but most of them involve rather awkward numbers. For instance (6·4), (9·8) qualifies, not to speak of a point which is approximately (8·535), (1·465).

You may notice a certain difference between the circle just drawn and the curve for $xy = 12$ which we had earlier. If we are looking for a point that passes the test $xy = 12$, and we think it would be a good idea to choose, say, 2 for x, then we have no choice for y; we must take 6 for y. But it is not so for the circle. As we have seen (2,1) and (2,9) both pass the test $x^2 + y^2 + 25 = 10x + 10y$. Even after deciding to choose 2 for x, we have more than one possibility for y. So the situation here is rather different from that in the games of Chapter 4, in which you call out a number and I answer. For here if you choose 2 for x, I may say

281

either 1 or 9 for y. On the other hand, $xy=12$ is exactly like those games. If you call out a number for x, I divide 12 by that number to find y. In shorthand, $y=12/x$.

It is naturally easiest to sketch graphs when we have such a definite rule. We usually make a table like this:

x	1	2	3	4	5	6
y	12	6	4	3	2·4	2

and read off the points that qualify, (1,12), (2,6) and so on. This method leads us directly to the points that pass the test and have to be marked in black. It does not mention explicitly the points that fail, but it is to be understood that all other points are to be marked yellow. For there is a definite rule that gives y when x has been fixed. This is the correct value, the only value that makes the point (x,y) pass the test.

Many books teach graphs by this method, and readers come to think of drawing graphs as a matter of making tables like the one just used. These readers may then find themselves puzzled when they meet a more complicated equation, such as the equation for a circle. The explanation, by which points that pass the test are marked black and those that fail are marked yellow, covers all possibilities. Starting from this idea, we can see why a table is

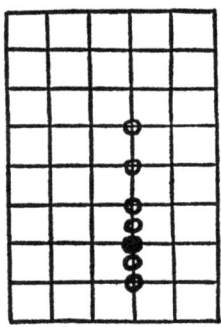

Figure 173

sometimes appropriate. With a definite rule like $y=12/x$ we may consider the points (2,3), (2,4), (2,5), (2,6), (2,7), $(2,5\frac{1}{2})$, $(2,6\frac{1}{2})$, and so on. For all of these, x, the first number, is 2. Only one of these points passes, (2,6), since $6=12/2$ is true. All the others fail, since 3 is not $12/2$ and 4 is not $12/2$, and so on. So we can mark our graph paper as in Figure 173. We get a vertical line, in which one point, (2,6) is black. All the others are yellow. If we then turn our attention to the points (3,1), (3,2), (3,3), (3,4), (3,5), etc. and mark the results in, we shall get another vertical line, containing one black point with all the rest yellow. We continue like this, getting many vertical lines. Each line contains one black point, and these black points lie on the graph of $y=12/x$. These black points are the ones that could be read off directly from the table for $y=12/x$. The table does in fact give the quickest and most practical way of drawing this graph. The discussion of black and yellow points is helpful for understanding what is meant by the graph of a more complicated equation.

Most beginners at graphs are asked to draw the graph $y=x^2$. From the table:

x	0	1	2	3
y	0	1	4	9

we see that the graph contains the points (0,0), (1,1), (2,4),(3,9). If we go on, we can see that the graph in Figure 174 also contains (4,16), (5,25), (6,36), and so on. The numbers for y get large very

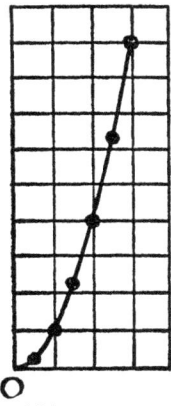

Figure 174

rapidly, so that a very tall, thin piece of graph paper is needed. Also the points just mentioned lie very far apart, and it is not easy to draw the curve through them. In Figure 174, instead of going on beyond (3,9) we have put in extra points corresponding to $x=\frac{1}{2}$, $x=1\frac{1}{2}$, and $x=2\frac{1}{2}$. These points are $(\frac{1}{2}, {}_{1})$, $(1\frac{1}{2},2\frac{1}{4})$, and $(2\frac{1}{2},6\frac{1}{4})$, and they help us to see how the curve goes.

Several mild surprises occur in the sketching of such curves, and these help to relieve the tedium of routine practice in making graphs. Suppose, for example, that after sketching $y=x^2$ we go on to $y=x^2+1-2x$. The table for this is:

x	0	$\frac{1}{2}$	1	$1\frac{1}{2}$	2	$2\frac{1}{2}$	3	$3\frac{1}{2}$	4
y	1	$\frac{1}{4}$	0	$\frac{1}{4}$	1	$2\frac{1}{4}$	4	$6\frac{1}{4}$	9

The table for $y=x^2$ was:

x	0	$\frac{1}{2}$	1	$1\frac{1}{2}$	2	$2\frac{1}{2}$	3
y	0	$\frac{1}{4}$	1	$2\frac{1}{4}$	4	$6\frac{3}{4}$	9

and we see a connexion between the two tables. If in the first table we disregard 1 and $\frac{1}{4}$, the first two entries for y, the remaining numbers in the y row are exactly the same as those in the table for $y=x^2$. How does this show itself in the graphs? (See Figure 175.)

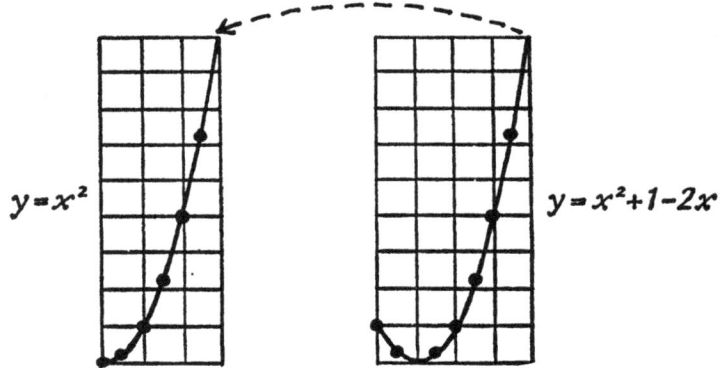

Figure 175

If the graphs are drawn on fairly thin paper, the children will find that they can hold the papers against the window, so that the graph of $y=x^2$ is exactly covered by part of the graph of $y=x^2+1-2x$.

Graphs

The equation $y=x^2+1-2x$ has been written in that order, rather than as $y=x^2-2x+1$, to avoid negative numbers coming into the calculation. For example, if we work out $y=x^2-2x+1$ for $x=1$, we get $1-2+1$, and some children may find difficulty in carrying out the instruction, 'Begin with 1. Take away 2 . . . '. Negative numbers do not come in at any stage of this calculation, so this work can be done before the idea of negative number has been reached.

The graph of $y=x^2+4-4x$ may be studied in this connexion. With this graph you can cover either of the graphs in Figure 175.

If you draw the graph of $y=10x-x^2$, taking values of x from 0 to 10, you get an arch like the curve in Figure 176. If you draw

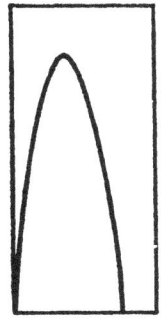

Figure 176

this graph and turn it upside down, you will find that part of it will exactly cover the curve $y=x^2$. In fact, all the four graphs we have considered recently are *the same curve in different positions.* Nor does it stop there. We can find many other equations that give this curve. We can make up a suitable equation in the following way. First think of any number, say 2. Then think of some other number, say 6, and write it with the letter x; this gives us $6x$. Finally take x^2. We now have 3 ingredients, 2, $6x$, and x^2. Join these together in any way you like by plus and minus signs. You might take x^2+6x+2 or $6x+2-x^2$ or x^2-6x-2, for example. You will find that $y=x^2+6x+2$ or $y=6x+2-x^2$ or $y=x^2-6x-2$ has a graph that can be placed to cover some part of the graph $y=x^2$. We may not get a very interesting graph. For

285

instance, if we take our first example above, $y = x^2 + 6x + 2$, this leads to the table:

x	0	1	2	3
y	2	9	18	29

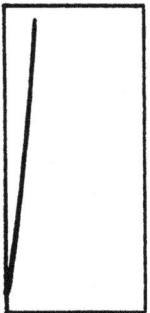

Figure 177

and its graph looks like Figure 177. You will find that it can be placed to cover the part of the graph of $y = x^2$ that goes through the points (3,9), (4,16), (5,25), and (6,36).

Sometimes you may get only a tiny piece of graph. For example $y = 3 - 2x - x^2$ gives us the curve in Figure 178 joining (0,3) to

Figure 178

(1,0). We hardly get a look at it before it is over the edge of the paper. And this raises a question which we shall consider in the next chapter; could we not perhaps bring along some more pieces of paper and follow the curve after it has left our original piece?

The curve that goes with $y=x^2$ and $y=10x-x^2$ and the other equations we have been considering is called a *parabola*. It occurs often in everyday life, for instance as the path of a ball thrown into the air on a calm day, the curve of the reflector on a car headlamp, the shape of the jet of water from a hose. This last example encourages us to believe that it may be possible to follow the curve beyond the edge of the paper, for a jet of water ceases to move in the shape of an arch when it hits an obstacle, but if the obstacle is removed, then the water continues in a smooth curve. Perhaps the edge of the paper represents an obstacle which we may succeed in removing. Children may find it stimulating to consider this idea for themselves. The graph $y=10x-x^2$ shown in Figure 176 is very suitable for discussion. Where do the children think this curve would go if we continued it to the right? Where would it go if continued to the left? The experiment indicated in Figure 175 may give them some ideas about where $y=x^2$ would go if it continued to the left.

Graphs naturally find a place in a book on vision in algebra. since graphs give us something to look at when we think about algebra. We do not learn graphs in order to draw neat graphs in an examination. We learn them rather to help ourselves to think about algebra. Whenever we meet an expression in algebra, we should think 'What would its graph look like?' and we should sketch a quick, rough little graph, just enough to show how the thing behaves. At first, these graphs will not tell us much, but as we keep drawing them, we shall find we get a feeling for them. There is a special reason for including graphs at this particular stage of this book. It is in order to ask the question posed a little earlier; can we follow a graph beyond the edge of the paper? The graph $y=10x-x^2$ reaches the edge of the paper when $x=10$, since $x=10$ makes $y=0$. So the graph seems to end at the point (10,0). What happens if we go beyond $x=10$ and consider $x=11$? When $x=11$, $y=110-121$ and we cannot deal with this without introducing negative numbers. If we say that $110-121$ is -11, we have to plot the point $(11, -11)$. Are we justified in doing this and where should this point be? These questions will be discussed in Chapter 12.

INVESTIGATING GRAPHS

As has already been mentioned, children should be encouraged to pose questions about graphs and to experiment freely. In the course of these experiments they may run into one or two difficulties. Curiously enough, difficulties tend to arise not with the most complicated but with the simplest situations. A thing can be so simple that it becomes, so to speak, invisible. For example, if a line is drawn through the points (0,2), (1,3), (2,4) children readily enough spot the law $y=x+2$. With the line through (0,1), (1,2), (2,3), they manage to spot $y=x+1$. But then they stick at the line shown in Figure 179 which goes through the

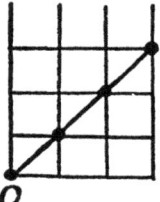

Figure 179

points (0,0), (1,1), (2,2), and (3,3). They notice surely that in each bracket the two numbers are the same, but they may come slowly to the law $y=x$ that expresses this fact. The reason may be that 'adding 2' and 'adding 1' are familiar operations in arithmetic, which they quickly recognize. 'Leaving alone' is not an operation that needs to be taught, so this idea may not come to a child's mind so readily. In terms of a guessing game $y=x$ corresponds to 'Whatever number you call out, I answer the same number'. In terms of a test to be passed, $y=x$ means that for each point, the two numbers in the bracket must be the same. Clearly (0,0) and (1,1) and (2,2) pass this test.

Another line that may cause difficulty is shown in Figure 180. The points on this line include (0,2), (1,2), (2,2), (3,2), and (4,2).

Graphs

What is there to be said about these? We could of course continue this line to get (5,2), (6,2), and so on. In fact we could get a point described by '(any number you like, 2)'. So to get on this line we can go any amount we like across, then 2 up. So if (x,y) is

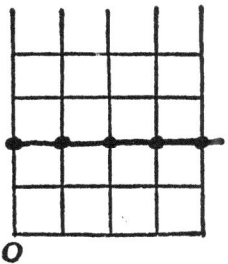

Figure 180

on this line, x may be anything at all, but y must be 2. What is the test, then, for being on the line? There is no point in looking at x, for x may be anything at all. We need only examine the second number, y. If y is 2, the point is on the line. If y is not 2, the point is not on the line. So $y=2$ is our test; $(17\frac{3}{4},2)$ is on the line, because the second number is 2; (3,4) is not on the line because 4 is not 2.

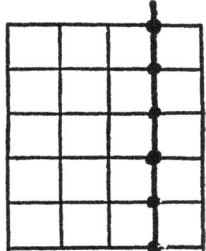

Figure 181

Similar remarks apply to the line in Figure 181. The points marked on this line are (3,0), (3,1), (3,2), (3,3), (3,4), and (3,5). You will be on this line if you start from 0, go 3 across, and any number up. The number across must be 3. So the equation for this line is $x=3$. To tell whether a point is on this line or not we only need to look at the first number, x, and see whether it is 3 or not. You may notice that this equation, $x=3$, will give great trouble to anyone who has been brought up by the rule. 'Make a table showing the values of y for $x=0$, $x=1$, $x=2$, and so on.' For we cannot make any such table; y is not even mentioned in the equation $x=3$. This is another example showing the advantage of the 'black and yellow' approach over the 'make a table' procedure.

CHAPTER TWELVE

Negative Numbers

MATHEMATICS grows gradually from one idea to another, and it is extremely difficult to say at what stage a new idea has been introduced. At the present moment it is hard to say whether this book has yet explained the idea of *negative number*. We have certainly been on the brink of it several times. For instance, in Chapter 10, we had pictures to illustrate the addition of $+2c$ and $-3c$. Yet we have never gone the whole way; we have never reached the stage where the instruction 'Think of a number' might receive the reply, 'I am thinking of -3'.

On page 251 we met 'the multiplication table of $+$ and $-$' but this did not involve the idea of negative numbers. Indeed it is remarkable that the multiplication table of plus and minus was known about two thousand years before negative numbers became generally accepted.* It was already known in antiquity that $n-3$ times $n-2$ was the same as n^2-5n+6, but this was always on the understanding that n was a number not less than 3. For example, n might be 10, as in our discussion on page 247, and the result that $10-3$ times $10-2$ is $100-50+6$ would have been quite familiar to a mathematician in ancient Babylon. It says that 7 times 8 is 56 and no novel idea is involved. It did not occur to anyone for centuries to wonder whether any sense could be made of taking 0 for n, so that $n-3$ times $n-2$ equals n^2-5n+6 would lead us to the idea that -2 times -3 should equal $+6$. Indeed this line of thought is highly unnatural. We pictured $10-3$ times $10-2$ as 10 rows, 3 of which were crossed out, each row containing 10 objects, 2 of which were crossed out. If we are to picture $0-3$ times $0-2$ in this way, we have to imagine that there are no rows to begin with, and that we cross out 3 rows; each row contains no object, and then 2 objects have to be removed from it. It is not surprising that this line of thought

* See Bell, *Development of Mathematics*, pp. 34, 62, and 96.

seemed unprofitable. Indeed, one might not merely think that it was incorrect to speak of negative numbers, but also that it was futile and pointless to bring them in. It is therefore worth while to consider an example which gives a slight indication that negative numbers can sometimes save us trouble.

There is a well-known type of problem, not of practical value in itself, but useful as an exercise in algebra. The problem might run: John is 10 years old and his father is 40. When will John's father be just twice as old as John? We start off, of course, by saying, 'Suppose it happens in x years time.' Then we have to think what the passing of x years does to the ages of John and his father. Suppose for example (though it is obviously not the answer to this question) that 7 years pass. John will then be 17 and his father 47. The age of each of them has increased by 7; 7 has been added to each of the numbers 10 and 40. Is there something special about 7? No; however many years pass, that number is added to their ages. So if x years pass, their ages will be $10+x$ and $40+x$. Children do not find it easy to analyse problems in this way, and it may be necessary to discuss this for quite a while with them, and to make a table something like the following:

	John's age	John's father's age
Now	10	40
3 years hence	13	43
7 years hence	17	47
x years hence	$10+x$	$40+x$

Now we go back to our problem; what does it ask us? It asks when John's father will be just twice as old as John. So we want to choose x to make this happen. Now x years hence, John's age will be $10+x$, so twice that will be $20+2x$. We want this to be the same as the father's age, $40+x$. That is, we want to choose x to make $20+2x=40+x$.

We may visualize this equation by the time-honoured picture of the balance; 20 and 2 bags are to balance 40 and 1 bag. If we remove a bag from each scale, this will not disturb the equilibrium. Then 20 and 1 bag will balance 40. It is fairly clear that the bag must contain 20. So $x = 20$, and in fact 20 years hence John will be 30 and his father 60, just twice as old as John.

The equation $20 + 2x = 40 + x$ could also be illustrated by rods, as in Figure 182. Here again it is pretty clear that x must be 20.

Figure 182

Suppose some children understand this method, and have practised it enough to be quite familiar with it. Then one day they meet the following question. In 1960, John was 25 and his father was 45. In what year is John's father just twice as old as John? It would be very natural for the children to start gaily, 'Suppose it happens x years after 1960'. Then John will be $25 + x$ and his father $45 + x$. Twice John's age will be $50 + 2x$, so that should be the same as his father's age, $45 + x$. So we write the equation:

$$50 + 2x = 45 + x$$

We imagine our balance, and take a bag off each side. This leaves us with:

$$50 + x = 45$$

We continue and take 45 off each side, which leaves us with $5 + x = 0$. At this point, if not earlier, we realize that the question is a little different from the one we had earlier. Clearly, however many years we wait after 1960, we shall not find a number x that makes $5 + x$ equal to 0. In fact, the longer we wait, the further we are from our objective. In 1975 John will be 40 and his father 60, so the father will be $1\frac{1}{2}$ times as old as John. If we wait until 2015, John will be 80 and his father 100, so the father will be $1\frac{1}{4}$ times as old. As $1\frac{1}{4}$ is further from 2 than $1\frac{1}{2}$, things are going the wrong way for us. One way to deal with this difficulty would be to say that we should have realized at the outset that we wanted a year before 1960 rather than after. We should on this view have begun by saying 'Consider the situation y years before 1960.' Then John was $25-y$ years old and his father $45-y$. We want to choose y so that the father's age is twice John's; that is, we want:

$$45-y=2\,(25-y)$$

This means: $\qquad 45-y=50-2y$

which we may picture as in Figure 183.

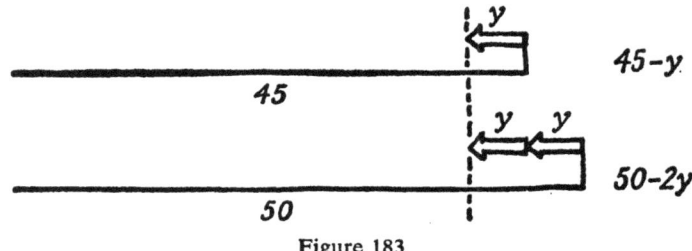

Figure 183

It is evident that $y=5$, so the desired year was 5 years before 1960, when John was 20 and his father 40.

At this point it might strike us that 5 came in our first equation. Our original attempt led to the equation $x+5=0$, which, if we allow the idea of negative numbers, would have the solution $x=-5$. Now we have found the actual solution to be '5 years earlier'. This suggests that we might be able to salvage something from our first attempt. We could interpret the result $x=-5$ as meaning: 'You asked the wrong question when you asked how many years *after* 1960; the answer, -5 *years after*, indicates that the correct answer is 5 *years earlier*.'

This way of dealing with the difficulty should appeal to pupils, for it means that we do not have to go right back to the beginning and work the whole problem through again. We perform the much easier and quicker task of interpreting the minus sign: -5 *years later* means 5 *years earlier*. In the same way, -3 *inches taller* would mean 3 *inches shorter*; -10 *degrees hotter* would mean 10 *degrees colder*; *a gain of* −8 *shillings* would mean *a loss of* 8 *shillings*.

The interesting thing to notice here is that at the time of our original argument we did not know that x was going to turn out to be a negative number, −5. We treated it as an ordinary number, and yet at the end we arrived at an answer which – if the interpretation given above is accepted – is correct. It is possible to prove mathematically that this will always happen. So far as the arguments and processes of algebra are concerned, negative numbers behave just like ordinary numbers. The proof of this assertion is not one that we can expect young children to understand. But children can gain confidence in negative numbers in the same way that the mathematicians of the seventeenth and eighteenth centuries did – by experimenting with them, by using them to solve a variety of problems, and finding that they give consistently correct results.

A teacher should not concentrate on one particular way of explaining negative numbers. An explanation that satisfies one child does not satisfy another. Nor should we set out to teach negative numbers on any one day. Rather, we should touch lightly on this subject, on many different occasions and in different ways. From all these encounters familiarity with negative numbers and confidence that their use is justified should gradually be built up.

Many children will have met negative numbers in connexion with temperatures below zero. This everyday experience helps to prepare them for the use of negative numbers in mathematics. In work on graphs we may write: 0, 1, 2, 3, as in Figure 184, to help

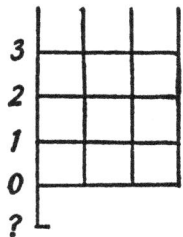

Figure 184

us when we are plotting points. Suppose in the course of some problem it became desirable to consider a point at the level of the question mark in Figure 184. What would be an appropriate label for this level? Children themselves without any prompting will often suggest −1.

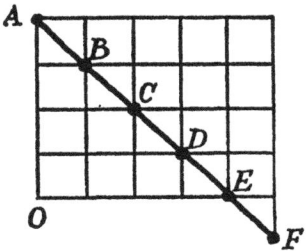

Figure 185

Graphs in fact have a natural tendency to run over the edge of the paper. In Figure 185, we have the points *A, B, C, D, E* in line. They are (0,4), (1,3), (2,2), (3,1), and (4,0), and they all pass the test $x+y=4$. But the line *ABCDE* runs naturally on to *F*. If we look at the first number in each of the brackets, we see 0, 1, 2, 3, and 4 for *A,B,C,D,* and *E*. It is easy enough to continue this sequence; for *F* we expect 5. And in fact to get from *O* to *F* we would begin by going 5 across. What about the second number in each bracket? These numbers are 4, 3, 2, 1, 0. The next number in sequence would be −1. This would lead us to describe *F* as (5,−1); that is, 5 across and −1 up. We are led to interpret '−1 up' as '1 down'. If we accept this interpretation, we can extend the line $x+y=4$ beyond the edges of our original squared paper. The point (5,−1) passes the test $x+y=4$, for we agreed in an earlier chapter that 5 and −1 add up to 4.

In the same way we can continue the line beyond *A* in the direction *EDCBA* to the point (−1, 5), where '−1 across' is understood to mean '1 to the left'.

Children can experiment with other lines such as $y=x+2$ or $y=x-3$ and see what happens when these are continued over the edge. Naturally, it is inconvenient to follow the lines over the actual edge of the graph paper. Rather we choose a point well inside the squared paper for O, and draw two lines to represent the edges. If children experiment in this way, they have a natural check on their accuracy. The line $y = x - 3$ looks as if, when continued, it would pass through the point $(2, -1)$. This checks and confirms the children's arithmetical work, for when we put 2 for x in $x - 3$ we get $2 - 3$ which is in fact -1. One may in fact follow this line to the points $(1, -2)$, $(0, -3)$, and $(-1, -4)$. This last point has both its numbers negative; it lies to the left and down from O.

When the pupils have arrived at and accepted the idea of using negative numbers on graph paper, they must of course be given enough practice with this to fix the idea in their minds. Such work can arise incidentally in the course of dealing with some other problem, and special exercises can also be devised. For instance, with the cat's face on page 163 a discussion might arise as to where the cat's paws should be drawn. Somewhere in the region of $(1, -3)$ and $(5, -3)$ might be suggested.

MULTIPLICATION OF NEGATIVES

It is the experience of many teachers that addition of positive and negative numbers can be taught quite easily along the lines suggested in Chapter 10, and subtraction with a little more difficulty. The serious difficulties were felt to arise when multiplication was reached, in particular with the maxim 'minus times minus is plus', so that, for example, -3 times -4 is $+12$. Here again the wisest treatment is to build up the pupil's faith in this result by a series of widely different experiences, in each of which it is found that minus times minus being plus gives the most reasonable or the most natural arrangement. The aim all the time is not to tell children the result, for the natural obstinacy and suspiciousness of human nature is such that the more we are told something, the less willing we are to believe it. Rather, we try to get the children to tell us; all the time we ask, 'What do you think should go in this space?' Children have a natural liking for order and pattern, and they will always try to give an answer that preserves and continues a pattern. In this respect, they have the same instinct as research mathematicians.

One approach is to show the multiplication table, set out as in Figure 186, and ask children whether they can extend it by filling

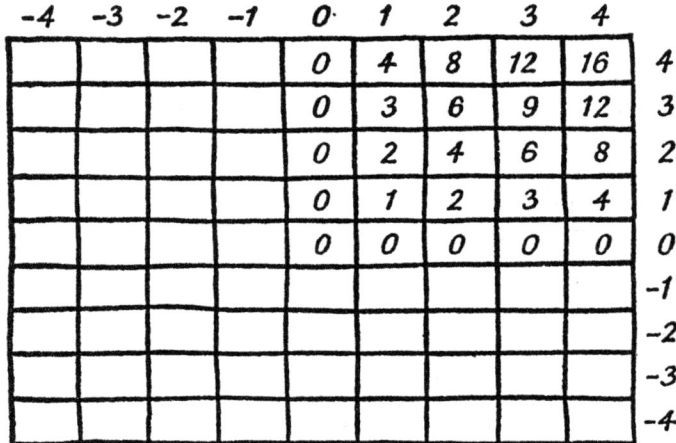

-4	-3	-2	-1	0·	1	2	3	4	
				0	4	8	12	16	4
				0	3	6	9	12	3
				0	2	4	6	8	2
				0	1	2	3	4	1
				0	0	0	0	0	0
									-1
									-2
									-3
									-4

Figure 186

in the empty spaces. At present, only positive numbers have been dealt with; for instance, where the column labelled 3 meets the row labelled 2 we find 6, which is 3×2. Most children readily extend the table. In the column labelled 0 they notice that every figure is 0, so they write noughts in all the spaces in this column. There is also a row containing only noughts. They extend this row in the same way. Then they may notice that the column labelled 1 reads downwards 4, 3, 2, 1, 0. The natural continuation is -1, -2, -3, -4, which they enter. The next column, reading downwards, gives 8, 6, 4, 2, 0, decreasing by steps of two. It is natural to continue -2, -4, -6, -8. Similar treatment will be given to the columns headed 3 and 4. Also the rows labelled 1, 2, 3, and 4 can be extended to the left in a similar manner. The table now appears as in Figure 187.

The children's trust in their guesswork will be increased if they observe that, where the column labelled 3 crosses the row labelled -4 we find -12, which we know already as a satisfactory answer for 3 times -4; 3 debts of £4 are a debt of £12. In the same

298

-4	-3	-2	-1	0	1	2	3	4	
-16	-12	-8	-4	0	4	8	12	16	4
-12	-9	-6	-3	0	3	6	9	12	3
-8	-6	-4	-2	0	2	4	6	8	2
-4	-3	-2	-1	0	1	2	3	4	1
0	0	0	0	0	0	0	0	0	0
				0	-1	-2	-3	-4	-1
				0	-2	-4	-6	-8	-2
				0	-3	-6	-9	-12	-3
				0	-4	-8	-12	-16	-4

Figure 187

way, in every other space we find the number that our earlier arguments have indicated ought to be there.

It is now easy to fill the remaining spaces. In the column headed -1 we see, reading downwards, -4, -3, -2, -1, 0. If these were temperature readings, they would indicate steady rises of $1°$, for $-3°$ is warmer than $-4°$. If this trend continues, we expect 1, 2, 3, 4 to follow, so we enter these in the remaining spaces of this column. In the same way, the column headed -4 contains -16, -12, -8, -4, 0, which represent repeated upward steps of 4. We expect the continuation 4, 8, 12, 16 and enter these numbers in the vacant spaces. At this point we could of course fill the columns headed -3 and -2 in the same way. However, it is rather more convincing to switch to the row labelled -3. Reading towards the left, we see -12, -9, -6, -3, 0, and this we continue with 3, 6, 9, 12. It will be found that the 3 and the 12 have already been written in, during our work on the columns headed -1 and -4. So we arrive at exactly the same numbers by continuing the rows to the left as we did by continuing the columns downwards. The jigsaw seems to be fitting together. And in the space for -4 times -3 the children have written the number 12, an 'ordinary' or positive number.

Vision in Elementary Mathematics

A logician may object that this is all guesswork and does not prove anything. We agree. It proves nothing. But it suggests a lot! If, using all kinds of different approaches, we find each time that 12 turns out to be the most natural result for -4 times -3, surely this cannot be pure accident. Something must be behind all these experimental results. That is all we are trying to convey to the children at present. Except with truly exceptional children, a vigorous mathematical and philosophical analysis should not be attempted until many years later. Indeed, even for the exceptionally intelligent and critical child, there is much to be said for our present approach, for it draws his attention to the fact that there is a problem here. We have shown that a certain result seems to work; we have not fully analysed why it works, nor have we proved beyond doubt that it always will give correct results. We have just given a few samples, a more than adequate basis for guesswork, but entirely insufficient for proof. Education should not merely provide completed information; it should leave children with some unanswered questions, which they can follow up for themselves later.

EVIDENCE FOR GRAPHS

Graphs give several useful ways of approaching negative numbers. If we draw the graph of $y=2x$, using positive numbers only for x and y, we obtain the line $OABC$ shown in Figure 188. At any point of this line, y is 2 times x. For example, C is (3,6) and the second number, 6, is just twice the first number, 3. We naturally want this test to continue to apply if we prolong the line beyond O. The line looks as if it goes through P, which is $(-1,-2)$, and Q, which is $(-2,-4)$. These points are specified by negative numbers, and we shall have little use for negative numbers if they tell us that P and Q do not lie on the line. But everything works out happily. P is $(-1,-2)$, and all our earlier arguments led us to believe that -2 is 2 times -1. In the same way, Q is $(-2,-4)$ and we have always thought that 2 times -2 is -4. So the system we have used does predict what in fact is true, that P and Q are on the line $OABC$ with the equation $y=2x$.

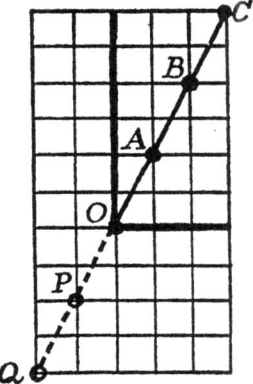

Figure 188

What happens if we try the rule that y is to be x times -2?
This would give us the table:

x	0	1	2	3
y	0	-2	-4	-6

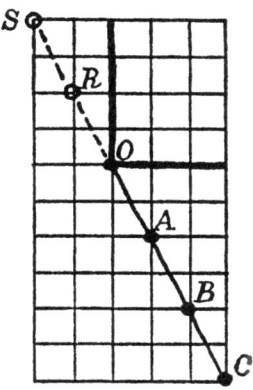

Figure 189

The graph will be the line *OABC* in Figure 189. If this line were
continued, it would pass through the points R and S. These
points should pass the test that y is x times -2. Now S is $(-2,4)$,

and has 4 for y and -2 for x. So 4 should be -2 times -2. So this confirms our belief that minus times minus should be plus. If we used any other result, S would fail the test. Now as S clearly is on the continuation of the line $CBAO$, this would mean that we would have to abandon our attempt to use negative numbers with graphs. We would have to keep to our original piece of paper, and conclude that the attempt to extend it was a failure, because wrong results followed, for example that S was not on the line. Either we have to accept minus times minus is plus, or give up the attempt to introduce negative numbers at all. In the same way, we may consider the point R, and see that -1 times -2 has to be 2 if things are to work out in a satisfactory way.

In Chapter 11, we saw that one and the same curve, in different positions and sometimes upside down, came from such different equations as $y=x^2$, $y=x^2+1-2x$, $y=x^2+4-4x$ and $y=10x-x^2$. If we take the curve for $y=x^2+4-4x$, which is shown in Figure 190: and place it over the graph of $y=x^2$, as in Figure 191,

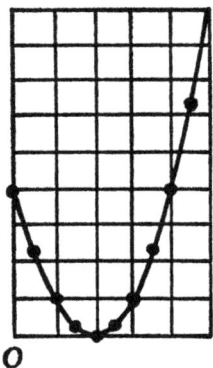

O

Figure 190

it suggests how that graph might most naturally be continued to the left.

It will be seen that the curve so obtained rises again as we pass to the left of O, and in fact passes through the points $(-\frac{1}{2},\frac{1}{4})$, $(-1,1)$, $(-1\frac{1}{2},2\frac{1}{4})$, and $(-2,4)$. Now for the graph of $y=x^2$ the

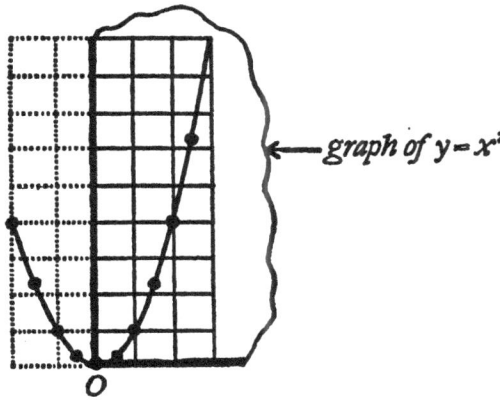

Figure 191

test is that the second number should be the square of the first. So we are led to believe that the square of $-\frac{1}{2}$ should be $\frac{1}{4}$, the square of -1 should be 1, and so on. 'The square of -1' means '-1 times -1', so once again we find we are led to the result minus times minus is plus.

Again we may experiment with a graph such as $y=x^2+5-6x$. The table for this reads as follows:

x	0	$\frac{1}{2}$	1	2	3	4	5	$5\frac{1}{2}$	6
y	5	$2\frac{1}{4}$	0	-3	-4	-3	0	$2\frac{1}{4}$	5

We have put in the values $\frac{1}{2}$ and $5\frac{1}{2}$ for x to help us in drawing these parts of the curve. If we did not accept the idea of negative numbers, we would have to omit everything between $x=1$ and $x=5$, and draw the graph as in Figure 192. But this is not very

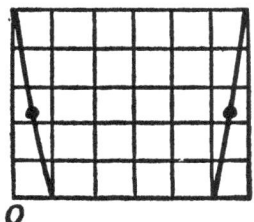

Figure 192

303

satisfactory. We feel that the graph dips below the edge of the paper and joins the two pieces of curve together. If we extend the graph paper, we can draw the curve as in Figure 193.

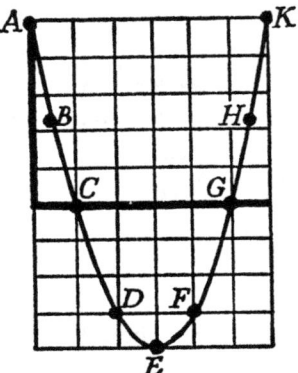

Figure 193

Now there is another way of looking at our equation, since x^2+5-6x is the answer obtained if we use the methods in Chapter 10 to find $x-1$ times $x-5$. So we may write x^2+5-6x $=(x-1)(x-5)$. Now in Chapter 10 we were thinking in terms of ordinary, positive numbers. In dealing with $x-1$ times $x-5$, we supposed x to be a number at least as big as 5. If we take 5 for x, then $(x-1)(x-5)$ gives 4×0 or 0 for y. So the graph of $y=(x-1)$ $(x-5)$ contains (5,0), which is H. In the same way, with 6 for x we get 5×1 for y, and $y=(x-1)(x-5)$ goes through (6,5), which is K. With $5\frac{1}{2}$ for x we get $4\frac{1}{2}\times\frac{1}{2}$, that is, $2\frac{1}{4}$ for y. The point $(5\frac{1}{2},2\frac{1}{4})$ is H. Thus, whether we start out to make a graph of $y=x^2+5-6x$ or $y=(x-1)(x-5)$ we get the points G, H, and K. In fact, whatever number larger than 5 we take for x, we find the same table for these two equations. Their graphs are exactly the same, so long as we are to the right of G. Now, if we are going to introduce negative numbers, and do our algebra by exactly the same procedures, not caring whether the letters stand for positive or negative numbers, we would like to arrange things so that the two graphs also coincide to the left of G.

Negative Numbers

The points D, E, and F lie below the edge of the original paper. For them we may make the following little table:

Point	D	E	F
Number for x	2	3	4
x^2+5-6x	-3	-4	-3
$(x-1)(x-5)$	$1\times(-3)$	$2\times(-2)$	$3\times(-1)$

The third row of this table shows the number you get for y if you use the equation $y=x^2+5-6x$. The fourth row gives the number found for y by using $y=(x-1)$ $(x-5)$. We want these two rows to agree, and this is, so to speak, not very controversial. Children do not find difficulty with the argument that 2 debts of £2 make a debt of £4, so $2\times(-2)=4$, and this is all we need to make the two results for E agree. With D and F also we use the same argument to see that 'plus times minus is minus'.

When we come to A, B, and C we reach the idea that surprises people most, 'minus times minus is plus'. The table for A, B, and C appears as follows:

Point	A	B	C
Number for x	0	$\frac{1}{2}$	1
x^2+5-6x	5	$2\frac{1}{4}$	0
$(x-1)(x-5)$	$(-1)\times(-5)$	$(-\frac{1}{2})\times(-4\frac{1}{2})$	$(0)\times(-4)$

Perhaps C is straightforward enough; 0 times anything should give 0. But for A we see that we can only get agreement if we accept that -1 times -5 is 5. Similarly for B, we want to have $-\frac{1}{2}$ times $-4\frac{1}{2}$ agree with $2\frac{1}{4}$. The size is right; half of $4\frac{1}{2}$ is $2\frac{1}{4}$. The sign, plus or minus, will be correct also, if we take minus times minus to be plus.

These examples illustrate something of the usefulness of negative numbers. If we accept the idea of negative numbers, and the multiplication table of plus and minus, we can specify the curve $ABCDEFGHK$ of Figure 193 by the equation $y=x^2+5-6x$ or $y=(x-1)$ $(x-5)$, whichever we prefer. The whole curve is given straight away by one equation. If we did not allow negative numbers, the equation $y=x^2+5-6x$ would give us the end pieces ABC and GHK, but the middle piece would be missing, and we would have to write a separate description of that. If we started with the equation $y=(x-1)(x-5)$ and refused to work with negative numbers, we would get only the piece *GHK*. We would probably need to write two further specifications – one to describe the part *CDEFG*, and one to describe the part *ABC*. Now the whole curve is one that hangs naturally together; it could arise in some problem about the path of a comet in the sky, or, as already mentioned, in a host of practical problems. It would be very inconvenient to have to specify it by two or three different descriptions. It is much easier to specify by one equation with negative numbers allowed, even though this means remembering that minus times minus is plus.

In fact, again and again in mathematics, using negative numbers allows us to join smoothly together two things that otherwise would have to be treated separately. For example, the 5 lines drawn in Figure 194 have the equations $y=x+2$, $y=x+1$,

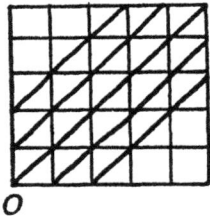

O

Figure 194

$y=x$, $y=x-1$, and $y=x-2$. In the first two we see an addition, in the last two a subtraction. So at first sight we might classify these lines in two families: in one family lines with equations '*y* equals *x* plus something', in the other '*y* equals *x* minus something'. In more formal terms, one family would contain lines with equations of the type $y=x+a$, the other those with equations of the type $y=x-b$. The line $y=x$ could go with either family; we could get it by taking 0 for *a* or 0 for *b*, since adding or subtracting nought makes no difference. This fact might make us wonder whether in fact the two families were really separate. On the graph paper, the 5 lines do not show any kind of separation; they look as if they all belonged to one kind. If we allow negative numbers we can bring this out, for subtracting 2 is the same as adding -2. We can say that all 5 lines are of the type $y=x+c$. If we replace c by 2 we get the first line, $y=x+2$. If we replace c by -2, we get the fifth line, $y=x-2$. For the other 3 we replace c by 1, 0, or -1.

In Chapter 11 we noted the fact that there are many equations which give the same curve as $y=x^2$ but in a different position. The equations $y=x^2-2x+3$, $y=x^2+5x-7$, $y=x^2+4x+11$, $y=x^2-9x-8$ are examples. If we allow negative numbers, we can cover all of these by saying $y=x^2+bx+c$. You can replace b and c by any numbers you like to choose, positive or negative. For example, to get $y=x^2-2x+3$ we choose -2 for b, $+3$ for c. To get $y=x^2+5x-7$ we choose $+5$ for b, -7 for c. Whatever numbers you choose for b and c, you will find you get a graph that will exactly cover the graph $y=x^2$. This graph will be 'the right way up', that is to say, it will look like a bowl, just as $y=x^2$ does. If instead you take $y=-x^2+bx+c$, you will find you can still cover $y=x^2$, whatever numbers you choose for b and c, but the graph will be 'upside down'; it will look like an arch rather than a bowl. For example, if in $y=-x^2+bx+c$ you can replace b by 10 and c by 0, you get the equation $y=-x^2+10x$. This equation, written in the form $y=10x-x^2$, we considered towards the end of Chapter 11, and saw that it appeared as an arch.

For the last three centuries it has been generally agreed that in algebra a letter may be replaced by a positive or a negative number. For example, suppose I say, 'Investigate graphs of the type $y=mx+c$'. I mean you are free to choose any number you like for m and any number you like for c. I want to know what you can find out about all the graphs that you might meet. In fact you would discover that, whatever numbers you choose for m and c, you would always get a straight-line graph. By saying 'the type $y=mx+c$' I thus convey to you that I am interested in knowing what can be said about a huge family of graphs. It would take me much longer to specify this family if you were not acquainted with the idea of negative numbers. For the family includes the following four examples:

(i) $y=2x+3$

(ii) $y=4x-5$

(iii) $y = -6x + 7$
(iv) $y = -8x - 9$

We can get them all from $y = mx + c$ by choosing positive and negative numbers, but to people who only know the numbers of elementary arithmetic these four seem quite unrelated, since they do not see that 'subtract 5' in example (ii) above is the same as 'add -5'. Example (iv) they would probably find meaningless, and example (iii) they would probably only accept if it were written in the form $y = 7 - 6x$.

With an understanding of negative numbers, we can use the single formula $y = mx + c$, and discover that it represents all kinds of lines situated in all kinds of different ways on the paper. Indeed the only lines whose equations cannot be covered by this form are the upright lines considered at the end of Chapter 11.

Please note in this chapter we have nowhere proved or attempted to prove that the graph of $y = mx + c$ is a straight line. The statement that every equation of this type is a straight line has simply been given without proof. Children can experiment by replacing m and c by different numbers, and they will find that they always get a straight-line graph. Indeed, it is instructive to try to classify these graphs. To do this, one should proceed in an orderly manner. For example, we might first investigate the cases in which m is replaced by 1. If we took c in turn as 2, 1, 0, -1, and -2, we should thus be led to the graphs of $y = x + 2$, $y = x + 1$, $y = x$, $y = x - 1$, and $y = x - 2$, the 5 lines already shown in Figure 194. We might then pass to the number 2 for m, and draw a figure showing the graphs of $y = 2x + 2$, $y = 2x + 1$, $y = 2x$, $y = 2x - 1$, and $y = 2x - 2$. In the same way we could draw five graphs to illustrate the effect of choosing 3 for m. Then we could do five more with -1 for m; these would be the graphs of $y = -x + 2$, $y = x + 1$, $y = -x$, $y = x - 1$, and $y = -x - 2$. We would continue in the same way with -2 for m, and then with -3 for m. Finally we might choose 0 for m. That is, we would consider $y = 0x + 2$, $y = 0x + 1$, $y = 0x$, $y = 0x - 1$, and $y = 0x - 2$. Really this is the simplest case of all, but pupils who are not familiar with multiplication by 0 may find it tricky. Of course, 0 times any number is 0, so the table for $y = 0x + 2$ would be this:

x	0	1	2	3
y	2	2	2	2

Its graph would be the top line in Figure 195. The other lines in this figure are the graphs of $y=0x+2$, $y=0x+1$, $y=0x$, $y=0x-1$, and $y=0x-2$.

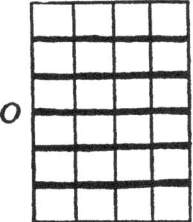

Figure 195

Pupils who carry out the investigation suggested above will find they have drawn the graphs of 35 lines, for they will have 7 diagrams (corresponding to 3, 2, 1, 0, -1, -2, -3, for m) with 5 lines on each (corresponding to 2, 1, 0, -1, -2 for c). By studying these diagrams they should be able to see how the choice of m affects the appearance of the graph. They can also see how changing the choice of c affects the graph. After some study of these 35 graphs, children can become so familiar with straight-line graphs that they can guess at sight the equation of a straight line drawn on graph paper, or, in the opposite direction, on being shown an equation they can immediately say what its graph will be.

THINGS AND UN-THINGS

Earlier we used several devices, such as mounds and hollows, credits and debts, things and un-things, to illustrate operations involving subtraction. In this earlier work, we usually kept within the limitation that there had to be more money than debts, more things than un-things. In this way, we did not at that stage need to discuss negative numbers, but we obviously came right to the edge of doing so. Having illustrated $10-3$ by 10 mounds and 3 hollows, why should we not omit the mounds, and illustrate −3 by 3 hollows? When we had 10−2 rows illustrated by 10 rows, followed by 2 rows which undid the effect of 2 earlier rows, why should we not delete from the diagram the first 10 rows, and be left with a picture of −2 rows?

We can in fact develop negative numbers in this way, and arrive at something very close to one of the justifications of negative numbers used by mathematical philosophers.

Suppose we are playing some game in which we may receive good marks or bad marks. A white counter is given to us for each good mark we earn. Every time we deserve a bad mark, we are forced to accept a red counter. It is understood that a red counter and a white counter cancel out. Someone with 13 white and 3 red counters has done just as well as someone with 10 white counters and no red counters; someone with 7 white and 17 red counters has done just as badly as someone with no white counter and 10 red ones.

When we come to speak of −3 we shall interpret the minus sign as meaning 'what wipes out'. Thus '−3 white counters' is interpreted as 'what wipes out 3 white counters', that is, '3 red counters'. In the same way, '−5 rows of 2 white counters' is interpreted as 'what wipes out 5 rows of 2 white counters', that is, '5 rows of 2 red counters'.

We then picture −5 times −3 by interpreting '−5 rows of −3 white counters'. We may interpret by stages. '−3 white counters' means '3 red counters'. So we are concerned with '−5 rows of 3 red counters', that is, with 'what wipes out 5 rows of 3 red counters'. This however is '5 rows of 3 white counters', making 15 white counters. We thus arrive at the result, −5 times −3 is 15.

It would also be possible to interpret '−5 rows of −3 white counters' by one rather long phrase, 'what wipes out 5 rows, each containing what wipes out 3 white counters'.

This approach, with white and red counters, seems in many ways the simplest and most clear-cut method for explaining negative numbers. However, as was mentioned earlier, we should not tie ourselves to any one method. A child's confidence in negative numbers should be built up by showing how all kinds of different arguments lead to the same conclusion.

Negative Numbers

It should be emphasized that our aim throughout has been a limited one. We have been trying to show that it is *reasonable* to believe that correct conclusions result from operations with negative numbers. We do not claim in any sense to have given mathematical proof of the validity of negative numbers. Indeed this is a task of some complexity. For example, it would require considerable mathematical and philosophical skill to start from our illustration of red and white counters, and show just why this gave correct answers about the ages of John and his father, and why it gave smooth continuations of the lines and curves on graph paper. Few children could fully appreciate the issues involved in such a discussion. Even fewer adults could explain the argument correctly. It is not wise to embark on a logical and philosophical analysis unless you and your pupil have the resources to carry it through. But quite young children can learn to appreciate that negative numbers do lead to results which can be seen to be satisfactory. They can become familiar with negative numbers as a working tool. If at a later age their curiosity or their scientific conscience leads them to demand some deeper analysis, they will have the background of knowledge that is necessary before any philosophical investigation can be fruitful.

CHAPTER THIRTEEN

Fractions

FRACTIONS are supposed to be difficult to learn. The reason for the difficulty is almost certainly a wrong approach in teaching. Too often the attempt is made to teach children a variety of complicated procedures without any real understanding, instead of teaching children to see the situation and decide for themselves what procedure is most reasonable.

The idea of a fraction is not a difficult one. If I hear that someone has left me $\frac{2}{3}$ of his estate, I understand the estate is to be broken into 3 equal parts, and I am to receive 2 of them.

One should never try to teach a child anything about operations on fractions without first making quite sure that the child understands what a fraction is. No elaborate apparatus is required. You can take a scrap of paper and ask the child to give you $\frac{2}{5}$ of it. You make it clear that this need not be done with extreme precision. If the child makes some attempt to split the paper into 5 equal portions, and to give you 2 of them, you know the essential idea is understood.

One of the first processes with fractions is 'cancelling'. The only clear idea this seems to produce in many children is that you cross out something on the top, and you cross out something on the bottom, and it makes no difference. This doctrine leads to extraordinary mistakes, very familiar to those who mark examination papers. There is no need for this confusion. The principle of cancelling is one which almost any child can think out for himself. We explain the basic idea of a fraction, that $\frac{3}{4}$ means that something is broken into 4 equal parts and 3 of them are given to us. We then set the children investigating, to see whether there might be several different ways of describing the same amount. For example, $\frac{1}{2}$ means that something is broken into 2 pieces and we are to have 1 of them. Is there any other way of specifying $\frac{1}{2}$? This investigation can be attacked in many ways.

Fractions

In one well-known method of teaching, children are asked to draw diagrams to illustrate various fractions, as in Figure 196.

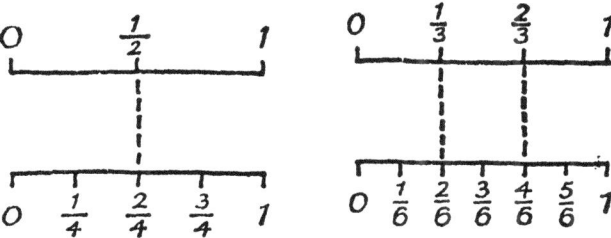

Figure 196

These diagrams resemble the graduations on a ruler. The children see that the marks for $\frac{1}{2}$ and $\frac{2}{4}$ come at the same place. They also notice that $\frac{1}{3}$ and $\frac{2}{6}$ represent the same position. There are many other coincidences of this kind. They can be encouraged to experiment and make guesses, with a view to being able to predict just when two fractions will coincide.

One might pose the question directly as a problem. A cake was cut into several pieces, and John ate several of them. In fact he ate half the cake. Discuss ways in which this might have happened. Perhaps there were 10 pieces and he ate 5 of them; this means he got $\frac{5}{10}$ of the cake.

One might stimulate the discussion by choosing various examples. What is half an hour? 30 minutes. What fraction is a minute of an hour? $\frac{1}{60}$, because to get a minute you have to break an hour up into sixty equal pieces. So half an hour is $\frac{30}{60}$ of an hour. In the same way, by considering half a foot or half a pound, one may reach the expressions $\frac{6}{12}$ or $\frac{8}{16}$ for a half.

Of course the discussion is not restricted to one half. Similar discussions and investigations are made with other simple fractions. It is found that all of them can be specified in many different ways. Now comes the interesting question; what kind of pattern, if any, lies behind all these different ways of expressing the same quantity?

There must be many different ways in which children could attack this problem. Some children will like to collect many

313

examples together and see what they can notice. They might make a list of all the ways they could discover of specifying a half.

They might then notice that it seems that you can have any number you like on top, but that the number below must be even, and in fact must be twice the number on top. It is extremely helpful if they are familiar with the very beginnings of algebra as described in Chapter 4. They can then write their conclusion very briefly. You can write x 'any number you like' on top, and you must write twice that number, $2x$, down below. So:

$$\frac{x}{2x}$$

gives all the possible ways of writing a half.

One could reach this conclusion in many other ways. John ate several slices, and in fact devoured half the cake. Could he have eaten 7 slices? Yes, if there were 14 slices in the whole cake. Could he have eaten 17 slices? Again, yes. It is in fact a very easy problem, if we are told how many slices John ate, to determine how many slices there were in the whole cake. We simply double the number. Could he have eaten any number at all? Yes, certainly. If he ate x slices, the cake contained $2x$ slices. So if we chop something into $2x$ pieces, and we receive x pieces, we have one half of the thing. This gives us the way of specifying a half displayed above.

The investigation of different specifications for the same fraction can proceed from diagrams. Could a quarter be expressed as 3 somethings? We draw our ruler, showing $\frac{1}{4}$, and below the line

we draw divisions to show that this quarter is 3 somethings. Can we complete this diagram to make a satisfactory picture? The children should tell us that we can. For the essential point about $\frac{1}{4}$ is that we divide the whole into 4 *equal* parts. If the first quarter contains 3 little pieces, each of the other quarters is the right size to hold 3 little pieces. Our completed diagram is thus as is shown overleaf.

It is now clear that the whole line has been divided into 12 pieces, so that each little piece is $\frac{1}{12}$. As $\frac{1}{4}$ contains 3 little pieces, $\frac{1}{4}$ must be the same as $\frac{3}{12}$.

Is there something special about 3 in this connexion? Could we do the same with any other number, say 5? We could. If the first quarter is divided into 5 parts, then each other quarter can be divided into 5 parts (all parts being the same size). The line is then divided into 20 parts, of which 5 make $\frac{1}{4}$. So $\frac{1}{4}$ is the same as $\frac{5}{20}$. It is clear that we could follow the same procedure with any number x. If the first quarter contains x little pieces, the whole line must contain $4x$ pieces. So, if we divide a thing into $4x$ pieces, and collect x of these pieces, we shall have a quarter of the original thing. As an equation:

$$\frac{1}{4} = \frac{x}{4x}$$

The argument that applies to 1 quarter can be applied to 3 quarters. In the diagram below, the arrow extends for $\frac{3}{4}$ of the line. If we divide each quarter into 5 smaller pieces, we obtain the

diagram below. There are now 20 little pieces in the whole line, and 15 of these lie under the arrow. The arrow is thus $\frac{15}{20}$ of the length of the whole line.

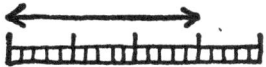

But, as it is $\frac{3}{4}$ of the length of the whole line, we must have

$$\frac{3}{4} = \frac{15}{20}$$

The children should consider this result. How have the larger numbers 15 and 20 come from the simpler numbers 3 and 4? It may strike them that $15 = 3 \times 5$ and $20 = 4 \times 5$; thus both the 3 and the 4 have been multiplied by 5. They may then discuss why this is so. Is it an accident, or could it have been foreseen? It is no accident, for each quarter of the line has been chopped into 5 smaller pieces, so that we have 5 times as many pieces under the arrow, and 5 times as many pieces in the whole line as we did before.

If we had divided each quarter into x pieces, there would be $3x$ under the arrow, and $4x$ in the whole line. The arrow originally covered 3 parts out of 4; now it covers $3x$ out of $4x$.

It is best if children are not told any rule, but given various exercises. Draw a picture to represent $\frac{2}{3}$ (below). If you divide

each third into 4 equal pieces, what new description does this give you for $\frac{2}{3}$? Answer: $\frac{8}{12}$.

As the children gain familiarity with this idea, one can gradually introduce numbers so large that drawing becomes tedious. Suppose we divide each third into 100 pieces. How many pieces will the whole line contain? 300. How many under the arrow? 200.

What new way will this give us of describing $\frac{2}{3}$? Answer, $\frac{200}{300}$, since the line is divided into 300 parts, and the arrow covers 200 of these.

By this and similar examples, the children pass from the stage of drawing to the stage of imagining. They can work out a series of questions. If in the picture for two thirds we divided each third into 7 pieces, what would be the new description of $\frac{2}{3}$ we obtained? What would we get if instead we divided each third into 11 pieces? In the course of reasoning and working out a series of such questions, the children would gradually become aware that each question involved multiplication.

$$\frac{2}{3} = \frac{200}{300} = \frac{2 \times 100}{3 \times 100}$$

$$\frac{2}{3} = \frac{14}{21} = \frac{2 \times 7}{3 \times 7}$$

Fractions

$$\frac{2}{3} = \frac{22}{33} = \frac{2 \times 11}{3 \times 11}$$

The brightest children would very quickly recognize the pattern, and would use this pattern as a short-cut. If you asked what 2 parts out of 3 became when each part was divided into 1,000 pieces, they would immediately shout out, '2,000 out of 3,000', without needing to go through the details of the argument. Other children would arrive at the same idea, but more slowly. Each child should be allowed to work at his own pace. Once a child has reached the stage where he can go straight to results such as those written above, he should not be forced to go through a longer argument based on drawings. On the other hand, no child should be urged to use the formal rule, 'you can multiply the top and bottom numbers by any number you like without altering the meaning of the fraction'; so long as he needs to think out what is happening, he should be encouraged to use pictures or apparatus, and spend as long a time as he wishes on the question. In time he will arrive at the formal rule for himself, and will have it securely, because it will be based on his own experience. Needless to say, whenever a child uses an incorrect method or has doubt about what he should do, we should go right back to the beginning and reason the whole thing out again. This may sound a slow procedure, but in fact it produces more knowledge in a given time than any other. Children who work without basic understanding invariably write nonsense. In examinations it becomes clear to the examiner that many children understand nothing of what they are doing; their correct answers are just as much accidents as their mistakes. Even for examination purposes, it is better to teach children to write a little sense than a lot of nonsense. But in fact a child who is taught early to think and to understand grows in confidence and knowledge as the years pass and reaches the stage where he can write a lot of sense in an examination, because he has thought out all the basic issues in advance.

In the work above, we began with a fraction in its simplest form, $\frac{2}{3}$, and obtained more complicated descriptions such as 200 three hundredths. Naturally, this process can be reversed, and we can work from the more complicated form to the simpler one.

What, for example, would be a simpler form for $\frac{4}{6}$? Figure 197

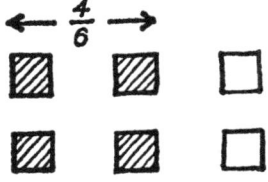

Figure 197

illustrates the meaning of the question. A piece of cardboard has been cut into 6 equal pieces, 4 of which are shaded. So $\frac{4}{6}$ of the cardboard is shaded. It is clear that some wasteful cutting has been done here. If we re-unite each piece in the lower row with the piece immediately above it we obtain the situation shown in Figure 198.

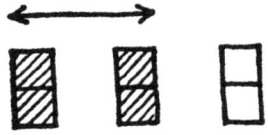

Figure 198

The cardboard is now in 3 pieces, 2 of which are shaded. So it would be simpler to say that $\frac{2}{3}$ of the cardboard is shaded. $\frac{2}{3}$ is thus a simpler specification of $\frac{4}{6}$.

It should of course be made clear to children that we are only concerned with the total amount. An unbroken vase is very different from a vase that has been broken into 4 identical pieces. So children might argue that 4 quarters is not the same as one whole. When we speak of fractions it is not implied that physical breaking up is actually carried out. Twin brothers could have half shares in a dog without sawing the animal in two.

The arrangement of pieces in Figures 197 and 198 suggests a way in which we can emphasize that *multiplication* is the operation involved when we replace a fraction by a more complicated specification. Figure 199 represents a piece of cardboard with $\frac{2}{3}$

Fractions

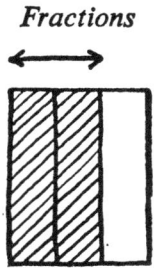

Figure 199

shaded. In Figure 200 we have the same piece of cardboard, with some level lines drawn on it.

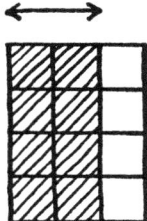

Figure 200

The cardboard is now divided into 12 pieces, of which 8 are shaded. The 8 shaded pieces form a rectangle, our standard picture for 2×4. The 12 pieces altogether form a rectangle, which we associate with 3×4. If pupils have learnt to associate multiplication with rectangles, this will help to remind them that:

$$\frac{2}{3} = \frac{2 \times 4}{3 \times 4}$$

It may help them to avoid the entirely incorrect view (easily accepted by a rote-taught child) that $\frac{2}{3}$ may be replaced by $\frac{2+4}{3+4}$!

OPERATIONS WITH FRACTIONS

Which is bigger, $\frac{2}{3}$ or $\frac{3}{4}$? My colleague, Robert Wirtz, has pointed out that this question (and other questions related to fractions) can be very simply investigated without any formal rules at all, by looking for a suitable object to consider. We want something that breaks easily into thirds or quarters. Suitable examples would be a foot length, a day, or a shilling. Suppose we take a foot, which is 12 inches. $\frac{2}{3}$ of a foot is 8 inches, $\frac{3}{4}$ is 9 inches. So $\frac{3}{4}$ is larger than $\frac{2}{3}$. We can, if we like, go on to say how much larger. For 9 inches exceeds 8 inches by 1 inch, and 1 inch is a twelfth of a foot. So we have in fact carried out a subtraction with fractions, for the result just found may be written:

$$\frac{3}{4} - \frac{2}{3} = \frac{1}{12}$$

Exactly the same device may be used for addition. How can we add $\frac{1}{4}$ and $\frac{1}{3}$? $\frac{1}{4}$ of a foot is 3 inches, $\frac{1}{3}$ of a foot is 4 inches. Together they make 7 inches, which is $\frac{7}{12}$ of a foot. We draw the conclusion:

$$\frac{1}{3} + \frac{1}{4} = \frac{7}{12}$$

It may happen that this method gives a slightly wasteful result. This does not cause any trouble. For example, some child might have worked this addition by using a day instead of a foot. $\frac{1}{3}$ of a day is 8 hours; $\frac{1}{4}$ of a day is 6 hours; the sum is 14 hours. As a day contains 24 hours, we have been led to express our answer in the form of 14 parts out of 24, instead of in the simpler form of 7 parts out of 12. However as 14 and 24 are both even numbers, it is fairly clear that the fraction $\frac{14}{24}$ can be put into a simpler form, by the considerations used in the first part of this chapter.

It is far better for children to work in this way by a method which they understand, even if occasionally a little extra work is involved. At a later stage, when fractions are thoroughly familiar, we may encourage the pupils to search for the most streamlined method. In this example, we might say, 'Yes, 24 can be split conveniently into thirds and quarters, but is there any smaller number that will do as well?' However, this type of criticism should come relatively late. Children should not be held back from exploring a method they have invented by the fear that it may not be the most efficient. In a class where different children invent different methods and argue about the reliability and convenience of these, there will be a far better understanding of logic, and a far better appreciation of the computational convenience of various approaches, than in a class where one stereotyped method (however streamlined) is prescribed and no child is encouraged to think or experiment for himself.

Even that bogey of arithmetic, division of fractions, seems to lose its terrors. The problem 'Divide $\frac{3}{4}$ by $\frac{2}{3}$' becomes 'How many times is 8 inches contained in 9 inches?' The answer is $1\frac{1}{8}$.

If we study the procedures followed in these examples, we find that they lie very close to the accepted formal methods. We say that $\frac{2}{3}$ of a foot is 8 inches. As an inch is $\frac{1}{12}$ of a foot, this is almost the same as saying that $\frac{2}{3}$ is $\frac{8}{12}$. In the same way, saying that $\frac{3}{4}$ of a foot is 9 inches is almost the same as saying that $\frac{3}{4}$ is $\frac{9}{12}$. And of course $\frac{9}{12}$ exceeds $\frac{8}{12}$ by $\frac{1}{12}$. An ordinary foot ruler provides the visual aid for this. Robert Wirtz's account of subtraction therefore follows essentially the same lines as the usual formal argument:

$$\frac{3}{4} - \frac{2}{3} = \frac{9}{12} - \frac{8}{12} = \frac{1}{12}$$

So one could begin to teach subtraction of fractions in Wirtz's way, and at a later stage (if this seemed desirable) teach the children to record the argument in the conventional form.

Exactly the same remarks apply to the example in addition, where a third and a quarter were added.

Wirtz's procedure for division is related to a formal procedure that is sometimes taught in schools and has certain advantages over 'invert and multiply'. To begin with, we write 'three quarters divided by two thirds' as a fraction:

$$\frac{\frac{3}{4}}{\frac{2}{3}}$$

We then use the formal principle stated in the first part of this chapter, that the meaning of a fraction is not altered if you multiply the top and bottom numbers by the same number. Here we choose to multiply by 12. The work thus proceeds:

$$\frac{\frac{3}{4}}{\frac{2}{3}} = \frac{\frac{3}{4} \times 12}{\frac{2}{3} \times 12} = \frac{9}{8} = 1\frac{1}{8}$$

One might accompany this written work by the spoken commentary – 'How many times is $\frac{2}{3}$ contained in $\frac{3}{4}$? How many times is $\frac{2}{3}$ of a foot contained in $\frac{3}{4}$ of a foot? How many times is $\frac{2}{3}$ of 12 inches contained in $\frac{3}{4}$ of 12 inches? How many times is 8 inches contained in 9 inches? $1\frac{1}{8}$ times.' This makes clear the connexion between the formal arithmetic and Wirtz's method of presenting division. Of course, the formal arithmetic can be understood without any such commentary by any pupil who understands the formal result, that we are permitted to multiply above and below by the same number, 12.

Some other ways of looking at division of fractions were given on page 100.

MULTIPLICATION OF FRACTIONS

A comment is desirable on multiplication of fractions. A language difficulty arises here. If you ask quite a young child, 'What is half of a half?' it is quite likely that he will be able to answer the question. He may take a cake, cut it into halves, then cut one half into 2 equal pieces, and know that 'a quarter' is the proper name for what he obtains. On the other hand, if he is given the written problem, 'Find $\frac{1}{2} \times \frac{1}{2}$', he may be completely stuck, and this even though he knows what the multiplication sign means in 2×3, and how fractions such as $\frac{1}{2}$ and $\frac{1}{4}$ are written.

In fact there is a very real intellectual difficulty in deciding just what multiplication should mean when fractions are involved. Many of the usual explanations and illustrations of multiplication fail completely when we cease to deal with whole numbers. For example, we may explain 3×4 by drawing 3 groups with 4 objects in each; we can hardly explain $\frac{1}{2} \times \frac{1}{2}$ by drawing half a group with half an object in each.

Yet, if we examine examples of multiplication in arithmetic, we shall find many that still make sense when fractions are involved. A gallon of water weighs 10 lb. What will 3 gallons weigh? The answer is 30 lb., since 30 is 3×10. We may equally well ask, what will $3\frac{1}{2}$ gallons weigh? The natural answer is 35 lb., since 35 is $3\frac{1}{2}\times10$. There might be some liquid, a little lighter than water, that weighed only $9\frac{1}{2}$ lb. a gallon. Then $3\frac{1}{2}\times9\frac{1}{2}$ would appear naturally when we calculated what $3\frac{1}{2}$ gallons of this liquid would weigh.

In the same way, we might have sticks 10 inches long; 3 such sticks would fill 3×10 inches. If the sticks were only $9\frac{1}{2}$ inches, 3 of them would fill $3\times9\frac{1}{2}$ inches. On the other hand, it is fairly clear what we mean by $3\frac{1}{2}$ sticks – 3 sticks, together with half a stick. Then $3\frac{1}{2}\times9\frac{1}{2}$ would have to be calculated, if we wanted to know how many inches $3\frac{1}{2}$ sticks would fill.

We can find a commercial illustration of $3\frac{1}{2}\times2\frac{1}{2}$. If a yard of material costs £2, to find the cost of 3 yards we need to know 3×2. But the cost might have been £2 10s. a yard, which is the same as £$2\frac{1}{2}$ and we might have required $3\frac{1}{2}$ yards. Here we have to work out $3\frac{1}{2}\times2\frac{1}{2}$. We can reason out the cost of $3\frac{1}{2}$ yards at £2 10s. a yard without any special method. We merely need to imagine actual money being put on the table. For the first yard, we put down two pound notes and a ten shilling note. We do the same for the second and the third. Finally, for the last half yard, we put down half as much – two ten shilling notes and five shillings. The collection of money on the table now appears as in Figure 201. The total amount here is £8 15s., or £$8\frac{3}{4}$. If we go back from money to pure arithmetic, we have the result that $3\frac{1}{2}\times2\frac{1}{2}=8\frac{3}{4}$.

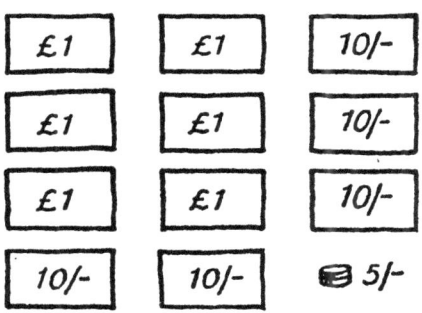

Figure 201

This picture can be put in another useful form. Imagine that instead of pound notes, we used a thin square plate of silver, and that this could be broken into 2 pieces whenever we wanted 10s., or into 4 when we required 5s. amounts. For precision, we suppose the square plates to be 1 inch by 1 inch. The collection of money in Figure 201 will now appear as in Figure 202.

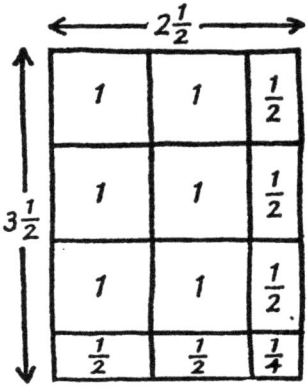

Figure 202

This figure will serve as an illustration of $3\frac{1}{2} \times 2\frac{1}{2}$. It will be seen that we have once more reached the rectangle as a picture of multiplication, but in a modified form. In our earlier work, we illustrated 3×4 by 3 rows of 4 objects. These objects were never broken into pieces. But now, with fractions, we use squares which can when necessary be broken into smaller parts.

If we wanted to work out $3\frac{1}{2} \times 2\frac{1}{2}$ by starting from this picture we should have to state the problem like this; draw a rectangle $3\frac{1}{2}$ inches by $2\frac{1}{2}$ inches; find how many tiles, 1 inch square, would be needed to cover it; that number is $3\frac{1}{2} \times 2\frac{1}{2}$. It is easy to count up the pieces shown in Figure 202 and to see that they add up to $8\frac{3}{4}$.

If we now return to our original problem of attaching a meaning to $\frac{1}{2} \times \frac{1}{2}$, it will be seen that we have several examples to bring out the meaning of this expression. It is related to all the following problems. If a gallon of liquid weighs $\frac{1}{2}$ lb., what is the weight of

$\frac{1}{2}$ gallon? (Answer, $\frac{1}{4}$ lb.) If a stick is $\frac{1}{2}$ inch long, how long is half the stick? (Answer, $\frac{1}{4}$ inch. Of course we assume the stick is sawn in half in the usual way – not split lengthwise!) If a yard of material costs £$\frac{1}{2}$, what will $\frac{1}{2}$ yard cost? (Answer £$\frac{1}{4}$.) If a rectangle is $\frac{1}{2}$ inch by $\frac{1}{2}$ inch, how many inch squares are needed to cover it? (Answer, $\frac{1}{4}$.) Figure 203 illustrates this result. The shaded area

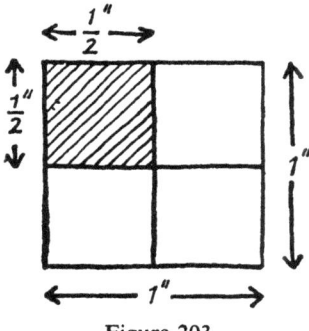

Figure 203

is $\frac{1}{2}$ an inch by $\frac{1}{2}$ an inch. It will be seen that 4 such pieces could be made out of a tile 1 inch square. So one quarter of a tile is needed to cover the shaded piece.

There is a language difficulty involved in multiplication of fractions. In everyday life we tend to say *times* when we are dealing with numbers larger than 1, and *of* when we are dealing with numbers less than 1. We say, 'This tree is $2\frac{1}{2}$ times as high as that' or 'Your holiday is $3\frac{1}{4}$ times as long as mine'. But we do not say 'This bush is $\frac{1}{2}$ times as high as that tree' or 'His holiday is $\frac{1}{4}$ times as long as mine'. But we do use such phrases as 'half of the height' and 'a quarter of the length'. Sometimes we do not even say *of*; we may say, 'The bush is half as high as the tree' or 'His holiday is only a quarter as long as mine'.

It is very natural that the word *times* should carry some suggestion of being plural, of involving more than one. This is the original, the primitive meaning of multiplication, 3 times 4. Half of 4 seems to have no connexion with it. Fractions have been used for thousands of years, and I do not know (perhaps nobody knows) their early history. I would be prepared to believe that

325

there was a time when people did not realize there was any connexion between *times* and *of*. Perhaps children had to learn two kinds of table, the 'times tables' and the 'of tables'. It may have been a revolutionary discovery when someone noticed that these two ideas fitted together, and that you could use the sign ×️ for both of them. This fitting together might have been noticed in many ways. Suppose you were asked to fill in the blanks in the following:

$$0, \ldots, 4, \ldots, 8, \ldots, 12$$

It would be natural to write 2, 6, and 10 in the spaces; you would then have the even numbers in order. If you were asked to fill the blanks in the following sequence:

$$0 \times 4, \ldots, 1 \times 4, \ldots, 2 \times 4, \ldots, 3 \times 4$$

the most natural answers would be $\frac{1}{2} \times 4$, $1\frac{1}{2} \times 4$, and $2\frac{1}{2} \times 4$. But the two problems are the same, for $0 \times 4 = 0$, $1 \times 4 = 4$, $2 \times 4 = 8$, and $3 \times 4 = 12$. This suggests that our answers also should be the same, that is, $\frac{1}{2} \times 4 = 2$, $1\frac{1}{2} \times 4 = 6$, $2\frac{1}{2} \times 4 = 10$. The last two sound reasonable when spoken, '$1\frac{1}{2}$ times 4 is 6, $2\frac{1}{2}$ times 4 is 10'. It is only the first one that sounds strange, '$\frac{1}{2}$ times 4 is 2'. The man in the street would say, 'Half of 4 is 2.'

Again, we might draw a picture to illustrate $3\frac{1}{2}$ times 4. We might imagine sticks 4 inches long, and draw 3 sticks and a half stick (Figure 204).

Figure 204

Our picture shows 3 fours and a 2. That is, it shows 3 *times* 4 and $\frac{1}{2}$ *of* 4, and these together make $3\frac{1}{2}$ *times*. Here is a strong suggestion that *times* and *of* are closely related.

Yet again, suppose we want to give a rule for finding the weight of any amount of liquid. If we have 3 gallons each weighing 10 lb., we know the weight is 30 lb. That is, we work out 3×10. So our rule would seem to be: write down the number of gallons and the weight in lb. of 1 gallon, and put a multiplication sign between. Now suppose we want to know the weight of $\frac{1}{2}$ gallon of some liquid, and that this liquid is so light that a gallon only weighs $\frac{1}{4}$ lb. According to our rule, we ought to work out $\frac{1}{2} \times \frac{1}{4}$. Now of course we know that $\frac{1}{2}$ gallon weighs half as much as a gallon, and that our answer ought to be $\frac{1}{8}$ lb., since this is half of $\frac{1}{4}$ lb.

We can deal with this situation in two ways. One way would be to stick to the customs of ordinary language. We would then have to have two separate rules – one for amounts more than a gallon, one for amounts less than a gallon. If a gallon of water weighed 10 lb., the rule for finding the weight of 3 gallons would tell us to calculate 3 *times* 10; the rule for finding the weight of $\frac{1}{2}$ gallon would tell us to calculate $\frac{1}{2}$ *of* 10. This would be quite a possible way of doing things. The other way is the one that mathematicians have found, more convenient and more satisfying, to use the single sign $\times$, with the understanding that 3×10 means 3 *times* 10 and $\frac{1}{2} \times 10$ means $\frac{1}{2}$ *of* 10. It is then found that everything fits together quite neatly.

The care needed in dealing with this question is due to the fact that immense philosophical questions are involved. A teacher can err in two opposite directions. Some teachers ignore the problem altogether: 'Don't you know that $\frac{1}{2} \times \frac{1}{2}$ means $\frac{1}{2}$ of a $\frac{1}{2}$?' 'You should be ashamed of yourself' is the implication. But a thoughtful child will see that some explanation is needed why one and the same sign, $\times$, should mean *times* on one occasion and *of* on another. Other teachers go to the other extreme. They raise all kinds of logical and philosophical questions for which most children are not ready. This is particularly likely to happen with a teacher who has just taken a course in philosophy or in modern mathematics. The trouble is that most philosophers see some aspects of life and are blind to others. Some for instance have maintained that mathematics is only an agreement to use language in a particular way. There is an obvious element of truth in this idea. Heaven did not provide man with stone tablets saying $+$ is to mean add and $\times$ multiply. The sign $\times$ was invented by some man, and presumably that man had the right to say what it meant. The philosophers stress that we may define a symbol in any way we like. So, if we liked, we could define $\frac{1}{2} \times \frac{1}{2}$ as 'a blood orange'. But philosophers tend to lose sight of the fact that nobody chooses to do so. The nature of the universe steers our thinking in a very definite direction. As has been shown by several arguments earlier, the simplest and most natural and most convenient interpretation of $\frac{1}{2} \times \frac{1}{2}$ is $\frac{1}{4}$.

Our teaching should try to steer a middle course. On the one hand, it should be clear that a problem exists in going from multiplication of whole numbers to multiplication of fractions. On the other hand, we should not subject young children to an abstract, logical analysis that does not correspond to their manner of thinking. Probably the best course is to pose the problem to them. We know what 3×4 means. Now we are starting on fractions; what ought $3\frac{1}{2} \times 4$ to mean? By a discussion of various practical problems, such as those earlier in this section, or by an illustration such as Figure 204, we bring the class to the conclusion $3\frac{1}{2} \times 4 = 14$. Then we may do $2\frac{1}{2} \times 4 = 10$ and $1\frac{1}{2} \times 4 = 6$. Finally we raise the question of $\frac{1}{2} \times 4$. The children themselves will probably volunteer the answer 2.

The contrariness of human nature is such that if you try to persuade a class that $\frac{1}{2}$ times 4 means $\frac{1}{2}$ of 4, they will not want to believe you. You should try to manoeuvre them into the situation where they are proposing that $\frac{1}{2} \times 4$ should mean 2, and you seem doubtful about this solution. Then they have to convince you that this is the best answer, and in the course of doing this they will also convince themselves.

If this procedure is followed, the children will have learnt two things from their experience, first, that there is a problem in defining multiplication of fractions, and second, that there is a natural solution of this problem. Now these are the two things that ought to be stressed in any philosophical discussion of the issue. A philosopher would discuss the problem explicitly. With children it is often better not to be too explicit. Summing up the experience in words often creates a feeling of puzzlement where the experience by itself would have left understanding.

Once it is clearly understood what $\frac{2}{3} \times \frac{4}{5}$ means, the problem of calculating this quantity is not difficult. If we visualize it with the help of a rectangle, as explained on page 324, we arrive at the picture in Figure 205.

The shaded area is $\frac{4}{5}$ inch by $\frac{2}{3}$ inch. It will be seen that the standard tile, 1 inch square, has been divided into 15 parts, of

Fractions

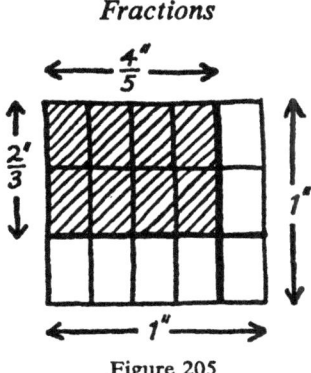

Figure 205

which 8 are shaded. The shaded area is thus $\frac{8}{15}$ of the tile. Children will notice that 8 is 2×4 and 15 is 3×5. In fact, the shaded area is a rectangle containing 2 rows of 4, and the whole tile shows 3 rows of 5. We thus reach the conclusion:

$$\frac{2}{3} \times \frac{4}{5} = \frac{2 \times 4}{3 \times 5}$$

In Figure 205 it will be noticed that a certain rectangle has been marked by a heavy line. This rectangle is $\frac{4}{5}$ of the whole tile. The shaded area is $\frac{2}{3}$ of this rectangle. So the shaded area is $\frac{2}{3}$ of $\frac{4}{5}$ of the tile. This gives us another way of seeing that the shaded area represents the fraction we want to calculate.

Wirtz's ideas could also be applied to another situation. We want to get rid of fractions, so we start by considering something that it is easy to find $\frac{2}{3}$ of $\frac{4}{5}$ of. Suppose we decide to take an hour, which is 60 minutes. $\frac{4}{5}$ of an hour is 48 minutes. $\frac{2}{3}$ of that is 32 minutes. So our answer is the fraction of an hour that 32 minutes is. This is $\frac{32}{60}$, which of course is not in its simplest form, and can be expressed more simply as $\frac{8}{15}$.

FRACTIONS AND ALGEBRA

Fractions and algebra interlock in several ways at different levels. In very early work with fractions, symbols such as x may give a convenient way of recording what children have learnt about

329

fractions. In fairly advanced algebra, children may need to re-member how fractions behave in arithmetic, in order to see what operations are justified in the manipulation of complicated algebraic fractions.

Earlier in this chapter we saw that 3 shares out of 4 were the same as 6 out of 8, or 30 out of 40, or in fact as 3 times any num-ber out of 4 times that number. It is much shorter to record this last phrase as $3x$ out of $4x$. Children can thus record explicitly in their notebooks the result of investigation into possible ways of writing $\frac{3}{4}$,

$$\frac{3}{4} = \frac{3x}{4x}$$

where x stands for any number we like to choose.

In the course of their work with fractions, children should frequently try to record, in the shorthand of algebra, the con-clusions they have reached.

At some stage they may want to find a suitable shorthand for 'any fraction'. A child may well argue that a fraction is written as 'any number you like to choose, divided by any number you like to choose'. As x stands for 'any number you like to choose', he will suggest the symbol $\frac{x}{x}$.

This is a very reasonable suggestion, but it overlooks the way we have always used the symbol x. It is true that x may stand for any number. But in the think-of-a-number tricks, once you had chosen a particular number, you had to stick to that number until you had finished the trick. If you began the trick with the number 3, when you came to the instruction, 'take away the number first thought of', you would not be allowed to subtract 5, on the grounds that you might have started with the number 5. You make a free choice at the outset, but you are not allowed to change your mind half-way through. An understanding of this kind holds throughout algebra. The fraction x/x means 'the number first thought of divided by the number first thought of'. If you thought of 3 it could mean 3/3; if you thought of 5 it could mean 5/5; but it could never mean 3/5. In fact, whatever number you choose for x, you will find that x/x equals 1. (It is understood that you are not choosing 0, for we do not know what 0 divided by 0 means.)

Children will agree that the two numbers in a fraction could be chosen by different people. Alf may choose the number at the top, and Betty the number at the bottom. The fraction may then be written a/b, where a stands for the number chosen by Alf, and b for that chosen by Betty.

With any fraction a/b we can do what we did with $\frac{3}{4}$; we can multiply the top and bottom numbers by any number x we like.

$$\frac{a}{b} = \frac{ax}{bx}$$

It might be necessary to remind the children that ax means a times x, and similarly for bx. This work is helping both to make children familiar with the notations of algebra and to bring out the facts of arithmetic.

When children are sufficiently familiar with any procedure in the arithmetic of fractions, they can be encouraged to express it algebraically. For example, earlier we multiplied $\frac{2}{3}$ by $\frac{4}{5}$. The same method would apply if the numbers 2, 3, 4, and 5 were replaced by others. So we suppose 2 replaced by a, 3 by b, 4 by c, and 5 by d – numbers chosen independently by Alf, Betty, Charles, and David. The children thus work from the particular result: $\dfrac{2}{3} \times \dfrac{4}{5} = \dfrac{2 \times 4}{3 \times 5}$

to its generalization:

$$\frac{a}{b} \times \frac{c}{d} = \frac{a \times c}{b \times d} = \frac{ac}{bd}$$

You will notice that we have first written $a \times c$ with the multiplication sign present, as is customary in arithmetic; later we have written ac, the form usual in algebra. Such little transitions are often necessary when children are beginning algebra, and have difficulty in remembering that ac means a times c.

If we wanted to add $\frac{2}{3}$ and $\frac{4}{5}$ by the usual formal method, we would first express $\frac{2}{3}$ as $\frac{10}{15}$ and $\frac{4}{5}$ as $\frac{12}{15}$. We would then argue that $\frac{10}{15}$ and $\frac{12}{15}$ make $\frac{22}{15}$. The essential step here is expressing both fractions as fifteenths. What makes us think of 15? The answer is easy; 15 is 3×5. We could apply the same method for adding a/b and c/d. We follow the working through, replacing 2 by a, 3 by b, 4 by c, and 5 by d, as before. In the arithmetic, our working began:

$$\frac{2}{3} = \frac{2 \times 5}{3 \times 5} \qquad \frac{4}{5} = \frac{3 \times 4}{3 \times 5}$$

The corresponding algebra would be:

$$\frac{a}{b} = \frac{ad}{bd} \qquad \frac{c}{d} = \frac{bc}{bd}$$

The final step was to add $\frac{10}{15}$ and $\frac{12}{15}$ to get $\frac{22}{15}$. We follow the same steps with the letters. The 10 is the top number in the first fraction; it corresponds to ad. The 12 is the top number in the second fraction; it corresponds to bc. We get 22 by adding 10 to 12; this corresponds to $ad + bc$. The number 15 is the bottom number both in $\frac{10}{15}$ and $\frac{12}{15}$, and it appears unaltered in the answer $\frac{22}{15}$.

So 15 corresponds to bd, and bd should appear unaltered at the bottom of the answer. So we reach the conclusion $\dfrac{a}{b} + \dfrac{c}{d} = \dfrac{ad + bc}{bd}$.

There is quite a bit of detailed thinking in this argument, and children need to take it slowly. We are guided partly by the places where the numbers occur. The pattern of the algebra is exactly like the pattern of the arithmetic. When we are in doubt what expression should replace a particular number, we see where it occurs in the arithmetical argument. We replace it by the letters that occur in that same place in the algebra. We are also guided by the sizes of the numbers. Our code runs rather as follows:

$$
\begin{array}{lll}
2 \text{ is replaced by} & a \\
3 & \text{by} & b \\
4 & \text{by} & c \\
5 & \text{by} & d
\end{array}
$$

10 is 2×5, so it is replaced by ad
12 is 3×4, so it is replaced by bc
15 is 3×5, so it is replaced by bd
22 is $10 + 12$, so it is replaced by $ad + bc$

It should be remembered that a stands for whatever number Alf has chosen, and so on. We can check whether we have found the correct argument by supposing that Alf, Betty, Charles, and Doris choose particular numbers, say 6, 10, 8, and 7. When we replace a, b, c, d by these numbers, the algebra should turn back into a sensible piece of arithmetic.

All the methods given in this chapter for the arithmetic of fractions can be translated into algebra.

For example, we can divide $\frac{2}{3}$ by $\frac{4}{5}$ using the following method:

$$\frac{\frac{2}{3}}{\frac{4}{5}} = \frac{\frac{2}{3} \times 15}{\frac{4}{5} \times 15} = \frac{10}{12}$$

If we use the code above, this becomes:

$$\frac{\frac{a}{b}}{\frac{c}{d}} = \frac{\frac{a}{b} \times bd}{\frac{c}{d} \times bd} = \frac{ad}{bc}$$

You may ask, 'Why did we not go on to simplify $\frac{10}{12}$ and express it as $\frac{5}{6}$?' The reason is that this cancelling was due to an accident in our original choice of numbers 2, 3, 4, 5. The numbers 2 and 4 are closely related. If we had chosen 2, 3, 5, and 7 for a, b, c, d, no cancelling would occur. This is a point that always has to be watched in algebra; to make sure that your argument applies for any numbers whatever, and that you have not been misled by some accidental feature of a particular case.

It may be noticed that this last result amounts to a statement of the 'invert and multiply' rule for division of fractions. We set out to divide a/b by c/d. Our answer is the same as we should have found if we had set out to multiply a/b by d/c.

How far this algebraic argument should be used in the teaching of elementary arithmetic is a matter for the judgement of the teacher. The deciding principle is one that applies to every use of algebra in arithmetic lessons. Algebra may be used as a shorthand provided the pupil is perfectly happy with this shorthand, and can use algebra confidently as a way of recording facts about arithmetic which he already understands. If a pupil is not clear about algebra, it will only increase his mystification if we try to teach him arithmetic through algebra.

FRACTIONS AND WHOLE NUMBERS

When a think-of-a-number trick is shown to a class of children, the numbers thought of are usually fairly small ones. When the trick has been found to work for all of these, some child may raise the question, 'Would the trick still work if I thought of a big number like 253?' It is found that the trick does work. Then some child may ask, 'Would it work for a fraction? What would happen if I thought of $3\frac{1}{2}$?' Again it is found that the trick works. Of course if the class do not raise these matters, the teacher can ask the children for their opinion on these questions.

In point of fact it can be proved that any trick that works for every whole number* also works for every fraction. This result will not be proved in this book. The proof is suitable for learners of algebra who are some way past the beginning of the subject. To make the assertion precise, it would be necessary to state exactly what we mean by 'any trick'. What limitations are we putting on tricks? We will suppose the trick asks a person to think of a number, and then work other numbers out from it, using as many additions, subtractions, and multiplications as the proposer cares to do.† A prediction is then made. If this prediction works whatever whole number may be chosen, then it will also work if any fraction whatever is chosen. For example, in various places we have mentioned that if you choose any number for n, and work out $(n+1)^2$ and also n^2+2n+1, these two results will turn out to be the same. Figure 94 on page 171 illustrated this result, and made us feel pretty certain that it would be true for any whole number. For this picture dealt with rows of dots, and we can draw any whole number of rows and any whole number of dots; we cannot draw fractions. But by the principle just stated,

---

* That is for 0, 1, 2, 3 . . . . We need not bother about negative numbers in this connexion, though the trick will work for them too.

† Often divisions are allowed also, but I have been cautious on the question of division as division by 0 might become involved and cause some complications.

Fractions

seeing the prediction holds when n represents any whole number, it also holds when n represents any fractional number whatever. In the same way, all the results justified by pictures with rows of dots will in fact remain true when n or x is no longer restricted to standing for some whole number. The picture will cease to apply but the result will in fact remain true.

MANIPULATION OF FRACTIONS

At a certain stage in the learning of algebra children meet problems such as that of simplifying the expression:

$$\frac{x+2}{x+3} - \frac{x}{x+1}$$

This is in fact a rather mild example of the sort of question that can be asked about fractions in algebra. This book has the limited aim of dealing with the beginnings of algebra, and detailed work on such fractions lies outside its scope. However, it seems worth while to take a brief look at the principles involved.

To handle fractions in algebra successfully two things are necessary – an understanding of fractions in arithmetic, and the ability to see the connexion between fractions in arithmetic and fractions in algebra. Mistakes arise when either of these is lacking.

The letters in algebra stand for numbers chosen at will. A step in algebra can be checked by putting particular numbers for the letters. If the check works whatever numbers may be selected, then the step is correct. For example, the step $\frac{a}{b} = \frac{ac}{bc}$ is correct. Whatever numbers we choose for a, b, c, we obtain a true statement in arithmetic (we tacitly assume that the letters b and c do not represent 0).

On the other hand the step $\frac{a+c}{b+c} = \frac{a}{b}$ (?) is not justified. One can find numbers for a, b, and c that make it true; for example, if we take a, b, and c all to be 2, we get $\frac{4}{4} = \frac{2}{2}$. But it is not true for all numbers. If we take 1 for a, 2 for b, and 1 for c, the equation becomes $\frac{2}{3} = \frac{1}{2}$, which is certainly false. In fact the equation makes the *false* assertion that a fraction is unaltered if you add the same number to the top and the bottom. Clearly only someone very familiar with fractions in arithmetic will be able to judge whether or not such general statements about fractions are true or false. Very often, of course, children are extremely hazy about the meaning of fractions in arithmetic; to embark on fractions in algebra is then a mockery of education. Even when pupils understand fractions in arithmetic, they do not always find it easy to see the connexion with algebra. An expression in algebra may be so long that children forget that it represents a single number. In arithmetic, a number is just a number, but in algebra we have, so to speak, a history of how we arrived at that number. Consider a very simple example of this effect. Suppose we are given the fraction:

$$\frac{a(c+d+e)}{b(c+d+e)}$$

Our usual collection of children, Alf, Betty, Charles, Doris, and Edward, are all choosing numbers. We have to add together the numbers chosen by Charles, Doris, and Edward. This sum, multiplied by Alf's number, is written on top. The same sum, multiplied by Betty's number, is written below. Suppose, for example, Alf chooses 2, Betty 3, Charles 4, Doris 5, and Edward 6. Then the sum $c+d+e$ is 15; $a(c+d+e)$ is 2×15 and $b(c+d+e)$ is 3×15. Our fraction then becomes:

$$\frac{2 \times 15}{3 \times 15}$$

We know that this is the same as $\frac{2}{3}$.

In arithmetic, we simply see 15 above. In algebra, we see $c+d+e$, which tells us (for the particular numbers used above) that this 15 was arrived at by working out $4+5+6$. For the purposes of cancelling, we are not so much interested in the history of the calculation. All we need to notice is that the number in brackets on the top, $c+d+e$, is the same as the number in brackets on the bottom, and that it enters into the fraction in such a way that cancelling is justified.

Fractions

With more complicated expressions the camouflage can be even more impressive. For example:

$$\frac{a(4cd^2e^2+7c^3e+d^4)}{b(4cd^2e^3+7c^3e+d^4)}$$

is the same fraction as a/b. Whatever numbers you choose for c, d, e, the long expression in the bracket gives the same number on top as it does below, so it cancels.

We now return to the expression mentioned at the beginning of this section:

$$\frac{x+2}{x+3} - \frac{x}{x+1}$$

Here, as usual, x stands for the number someone has thought of. If that number happened to be 2, the expression would signify:

$$\frac{4}{5} - \frac{2}{3}$$

But of course we do not know that 2 is going to be thought of. If some other number were chosen for x, we should get other fractions; for example, if 10 were chosen, we would have:

$$\frac{12}{13} - \frac{10}{11}$$

You may notice that in each case 4 numbers are involved that follow each other when we count – 2, 3, 4, 5 in the first instance, 10, 11, 12, 13 in the second.

We can get some guidance for the algebra by thinking what we should do with the arithmetic in the special cases. In the first example, we would express $\frac{4}{5}$ as $\frac{12}{15}$ and $\frac{2}{3}$ as $\frac{10}{15}$. How are we led to 15 here? Very simply, 15 is 5×3. In the same way, in the second example, we would try to express both fractions in such a way that 13×11 came at the bottom of both.

In both these examples, the number required is found by multiplying together the numbers at the bottom of the two fractions. This gives us the clue for what we should do with the fractions in algebra. Of course the trouble in the algebra is that

337

the numbers at the bottom are not nice, definite numbers like 5 and 3 or 13 and 11. They are $x+3$ and $x+1$; that is, they are given by a prescription. Someone is to think of a number; 3 more than that number and 1 more than that number are to be written at the bottom in the two fractions. It may be necessary for pupils to carry out the arithmetic of a whole series of particular cases until they see that a common procedure is being followed in all of them, which can be stated in the language of algebra.*

In our first arithmetical example we supposed x, the number thought of, to be 2. Then 3 came in as $x+1$, 4 as $x+2$, and 5 as $x+3$. This may help us to work back from the particular case in arithmetic to the general argument of algebra. We try the effect of replacing 2 by x, 3 by $x+1$, 4 by $x+2$, and 5 by $x+3$. Care is needed in doing this to see that we are not led astray by accidental connexions between the numbers. For example, 4 is the square of 2, but $x+2$ is not usually the square of x.

In arithmetic our first step was : $\dfrac{4}{5} = \dfrac{4 \times 3}{5 \times 3}$

This suggests that in algebra our first step should be:

$$\frac{x+2}{x+3} = \frac{(x+2)(x+1)}{(x+3)(x+1)}$$

We see that this step is justified, for the meaning of a fraction is unaltered if the top and bottom numbers are multiplied by the same number. And that 'same number' could be $x+1$; that is, one more than the number somebody has thought of.

Again, in the arithmetic we carried out a similar re-writing of the other fraction.

$$\frac{2}{3} = \frac{5 \times 2}{5 \times 3}$$

---

* In this particular example, it would be wise to use only even numbers for x. When x is odd, there is an effect of cancelling in the arithmetic, which obscures the pattern. If we take 1, 2, 3, 4 in turn for x, we find that the expression being studied takes the values $\dfrac{1}{4}, \dfrac{2}{15}, \dfrac{1}{12}, \dfrac{2}{35}$. To see the pattern, we need to express these in the form $\dfrac{2}{8}, \dfrac{2}{15}, \dfrac{2}{24}, \dfrac{2}{35}$. We now have 2 on top in each case. At the bottom we have 2×4, 3×5, 4×6, 5×7.

Fractions

This suggests the step in algebra which we see is justified:

$$\frac{x}{x+1} = \frac{(x+3)x}{(x+3)(x+1)}$$

If any pupil does not accept this argument, or does not see what is being done, he could be advised to choose some particular number for x, and see whether the statements make sense for that number. So long as he feels doubt, he should continue to try further numbers for x. In the course of doing this he will come to see the meaning of the algebra and (provided always that his knowledge of fractions in arithmetic is adequate) to see that the steps are justified.

The stage now reached corresponds in arithmetic to seeing that subtracting $\frac{2}{3}$ from $\frac{4}{5}$ is the same problem as subtracting $\frac{10}{15}$ from $\frac{12}{15}$. Corresponding to a *fifteenth* we have the rather lengthy $(x+3)(x+1)$th. Instead of trying to carry this awkward expression around, let us call this quantity a *piece* for short. The two fractions involved in the subtraction have now both been expressed as so many pieces. The first fraction is $(x+2)(x+1)$ pieces, the second is $(x+3)x$ pieces. If we work these out, the first is x^2+3x+2 pieces, the second is x^2+3x pieces. Subtracting, we see that the difference is 2 pieces, or, in full, two $(x+3)(x+1)$ths. So we may write:

$$\frac{x+2}{x+3} - \frac{x}{x+1} = \frac{2}{(x+3)(x+1)}$$

This result may be checked for accuracy by trying different values of x in it. It should be compared with the fractions given in the footnote on page 338.

ACADEMIC PROBLEMS WITH FRACTIONS

Someone has said that intellectual ability consists in the power to hold many ideas in mind and to see the relationships between them. It has also been said that pupils of limited ability find it' hard to keep one idea in mind and impossible to keep two.

Manipulation of fractions in algebra, except perhaps in the very simplest cases, seems to call for some intellectual power.

Vision in Elementary Mathematics

Many ideas have to be held together. First of all, the whole theory of fractions in arithmetic must be known thoroughly. The student of algebra must be able to recognize immediately whether a particular operation with fractions in arithmetic is legitimate or not. He must be able to link this knowledge to algebraic expressions, which do not represent any particular fraction, but can represent one of a large number of fractions, depending on what x, the number thought of, may happen to be.

For the less able children, the manipulation of fractions in algebra is probably not a very profitable study. There are plenty of topics in algebra which are good fun and are intellectually stimulating to all children. Some of these, it is hoped, may have been indicated in the earlier chapters of this book.

For children of higher ability the situation is less clear. From time to time, mathematicians say that schools spend too much time on 'mere manipulation'. But they also complain if students come to university and are unable to deal readily with operations in algebra. The truth seems to be that algebraic manipulation plays a small part in some branches of mathematics but is still important in others, as it also is in science and engineering. It is wrong to make pupils spend hours on repetitive and soul-destroying drill. It is still desirable that pupils should spend sufficient time working at interesting and stimulating problems involving algebra for them to become thoroughly familiar with algebra and able to follow an algebraic argument in a textbook or scientific paper without strain or undue expenditure of time.

Answers

2. Type $4n+3$.

3. 3, 13, 23, etc. All such numbers end in 3.

4. Remainder is always 1.

5. 0.

6. Remainder, 1. Type $3n+1$.

7. 1.

8. 1.

10. The numbers in the *Drew* column add up to 9. But as every draw involves *two* teams, this total ought to be an even number.

1. 20_5; 9.

2. 13_7; 30.

3. 1, 10, 11, 100, 101, 110, 111, 1,000, 1,001, 1,010.

4. $10,000_2$.

5. $222_3 = 26$.

6. 46; even.

7. 1_3 and 111_3 are odd, the others even.

8. $3 \times 1 = 3$; $3 \times 2 = 10_6$; $3 \times 3 = 13_6$; $3 \times 4 = 20_6$; $3 \times 5 = 23_6$.
(The numbers in this table end in 0 or 3 just as, in our usual system, the numbers of the 5 times table end in 0 or 5.)

9. The numbers in the tables are: 1 times, 1, 2, 3, 4,; 2 times, 2, 4, 11_5, 13_5; 3 times, 3, 11_5, 14_5, 22_5; 4 times, 4, 13_5, 22_5, 31_5.
In the 4 times table the digits add up to 4.

10. $11_6 \times 11_6 = 121_6$.

11. $11_3 \times 11_3 = 121_3$; $11_4 \times 11_4 = 121_4$;
$11_5 \times 11_5 = 121_5$; $11_7 \times 11_7 = 121_7$;
$11_8 \times 11_8 = 121_8$.
In bases 3, 4, 5, 6, 7, 8 we have $11 \times 11 = 121$, just as in our usual system.

12. Each result resembles $12 \times 12 = 144$.

13. Each result resembles $12 \times 13 = 156$.

14. Even. In base 8, as in base 10, a number is even when the last digit is even.

15. (i) The number is even when the sum of the digits is even. (ii) The number divides by 3 when the sum of digits divides by 3. For $1,111_7$ the sum of digits is 4. This number is even but not divisible by 3.

Pages 49, 50

1. $m=9, s=1.$
2. $m=5, s=3.$
3. $m=5, s=2.$
4. $m=5\frac{1}{2}, s=2\frac{1}{2}.$
5. $m=6, s=2.$
6. $m=5, s=2.$
7. $m=6, s=2.$

Page 56

1. $m+3s, m-2s$, difference $5s$.
2. $2s$.
3. $5s$.
4. $10s$.
5. $28s$.
6. $m-2s, m-5s$, difference $3s$.
7. $m-s$ is larger than $m-3s$ by $2s$.

Page 75

$x; x+1; x-1; 2x; 3x; 3+x$ or
$x+3; 2x+1$ or $1+2x; 10-x; x-5$

Page 79

1. $y=x+2.$ **2.** $y=10-x.$ **3.** $y=2x.$ **4.** $y=10x.$
5. $y=10x+1.$ **6.** $y=2x+1.$ **7.** $y=\frac{1}{2}x.$

Pages 141–4

1. (i) About 126 miles.
(ii) About 155 miles.
(iii) About 179 miles.
(iv) About 900 miles.

Answers

2. (i) 5 inches; (ii) 5 inches; (iii) 5 inches; (iv) 5 inches.
3. 15 feet.
4. $s^2 = 65$, $t^2 = 65$. The lines are exactly the same length.
5. 9·95 inches, very nearly.
6. $s^2 = 50$, $t^2 = 50$. Exactly equal.
7. (i) Both 5; (ii) Both 5; (iii) Equal to s and t in question 6.
DEF would just cover *ABC*

Pages 149, 150
1. $-$, $+$ in each case.
2. To add 1.
3. To add 2.
4. To add 10.
5. Leaves it unaltered.
6. Subtracts number from 20.
7. Subtracts number from 80.
8. Adds 5.
9. Question 4; $10+x$.
10. Question 3; $2+x$.
11. Question 5; x.
12. Left to reader.

Page 153.
1. 7, 9, 11.
2. 10, 20, 30.
3. 1, 2, 3.
4. 5, 15, 75.
5. 1, 10, 11.

Page 157.
1. 4, 7, 5, 8, 6.
2. 6, 4, 6, 4, 4.
3. 15, 15, 9, 31, 5.
4. 15, 25, 25, 35, 20.
5. $1\frac{1}{2}$, $1\frac{1}{2}$, $\frac{1}{2}$, $3\frac{1}{2}$, $\frac{1}{2}$.

Pages 157, 158.

Left as investigation for reader.

Pages 171, 172

1. (i) 1,002,001. (ii) 100,020,001. (iii) 10,000,200,001.
(iv) 1,000,002,000,001. (v) 40,401. (vi) 4,004,001.
(vii) 9,006,001. (viii) 90,601. (ix) 961. (x) 441.

2. 100^2 exceeds 99^2 by 199, so $99^2 = 9,801$.

3. $13^2 = 169$, $31^2 = 961$.

4. Investigation leads to conclusion $(n+a)^2 = n^2 + 2an + a^2$.

5. $n + (n-1) = 2n - 1$.

6. $(n-1)^2 = n^2 - (2n-1) = n^2 - 2n + 1$.

Page 177

A.1. $(n+1)(n+3) - n(n+4) = 3$.
A.2. $(n+1)(n+4) - n(n+5) = 4$.
A.3. $(n+2)(n+3) - n(n+5) = 6$.
A.4. $(n+1)(n+1) - n(n+2) = 1$.
A.5. $(n+2)(n+2) - n(n+4) = 4$.
B.1. $n(n+4) - (n+1)(n+2) = n - 2$.
B.2. $n(n+5) - (n+1)(n+3) = n - 3$.
B.3. $n(n+5) - (n+2)(n+2) = n - 4$.
B.4. $(n+2)(n+3) - n(n+4) = n + 6$.
C.1. $(n+1)(n+3) - n(n+2) = 2n + 3$.
C.2. $(n+1)(n+4) - n(n+3) = 2n + 4$.
C.3. $(n+1)(n+5) - n(n+4) = 2n + 5$.
D.1. $(n+2)(n+2) - n(n+1) = 3n + 4$.
D.2. $(n+1)(n+4) - n(n+2) = 3n + 4$.

Page 180

1. $n^2 + 6n + 8$.
2. $n^2 + 7n + 10$.
3. $n^2 + 7n + 12$.
4. $n^2 + 8n + 15$.
5. $n^2 + 4n + 3$.
6. $n^2 + 5n + 4$.

Page 183

1. $(n+3)(n+4)$.
2. $(n+2)(n+6)$.

3. $(n+1)(n+12)$.
4. $(n+2)(n+4)$.
5. $(n+3)(n+5)$.
6. $(n+1)(n+3)$.
7. $(n+1)(n+5)$.
8. $(n+1)(n+7)$.
9. $(n+1)(n+24)$.
10. $(n+2)(n+12)$.
11. $(n+3)(n+8)$.
12. $(n+4)(n+6)$.
13. $(n+1)(n+1)$.
14. $(n+2)(n+2)$.
15. $(n+3)(n+3)$.

Page 185

1. $(n+3)(n+24)$.
2. $(n+25)(n+36)$.
3. $(n+12)(n+75)$.
4. $(n+28)(n+30)$.

Pages 219, 220

1. $x+1$ rows of x dots; 1 left over. $x+1$, remainder 1.
2. $x+2$ rows, none left over. The picture is Figure 53 on page 115.
3. 2 rows, 3 over.
4. $x+1$. See Figure 94 on page 171.
5. x rows, 5 over.
6. x, with remainder 2.
7. (i) $9=4\times2+1$.
(ii) $36=3\times10+6$.
(iii) $1057=10\times100+57$.
8. (i) $2x+5=2(x+1)+3$.
(ii) $x^2+x+1=x(x+1)+1$.
(iii) $x^4+x^2+x+1=x^2(x^2+1)+x+1$.
9. $x^2+5x+20=(x+2)(x+3)+14$.
10. $x^4+2x^3+3x^2+2x+1$; 12,321.
If we put 10 for x in $(x^2+x+1)^2$ we get 111^2.
11. $x^4+4x^3+8x^2+8x+4$; 14,884.
Same numbers occur in each.

345

12. $x^4+4x^3+10x^2+12x+9$; 15,129.

Clearly, numbers as large as 10 and 12 cannot appear as digits. The carrying in 123^2 causes the result in arithmetic to appear different from the result in algebra.

13. (i) $4x^2+4x+1$. (ii) $4x^4+8x^3+4x^2$.

(iii) $4x^4+8x^3+8x^2+4x+1$.

The triangle would be right-angled, for it satisfies Pythagoras's condition (Chapter 6).

Pages 259–62

1. $+6$.

2. -2; $10+3$ and $10-2$; correct.

3. x^2+x-6. Replacing x by 10 would make the numbers as in question (2).

4. $(x-2)(x+2)$; x^2-4.

5. Entries in square are x^2; $+3x$; $-3x$; -9.

Answer, x^2-9.

6. x^2-16; x^2-25; x^2-a^2.

7. In square x^2; $+ax$; $-ax$; $-a^2$.

Answer x^2-a^2, as guessed in question (6).

8. x^2-1. Check, $11 \times 9 = 100-1$.

9. x^3-1. Check, $111 \times 9 = 1,000-1$.

10. x^4-1, $111 \times 9 = 10,000-1$.

11. $(x-1)$ $(x^4+x^3+x^2+x+1)=x^5-1$ is suggested and is correct.

12. Yes; $x^5+x^4+x^3+x^2+x+1$.

13. In square, x^4, x^3, x^2; x^3, x^2, x; x^2, x, 1.

Answer, $x^4+2x^3+3x^2+2x+1$.

14. In square, $+x^4$, $-x^3$, $+x^2$; $-x^3$, $+x^2$, $-x$; $+x^2$, $-x$, $+1$.

Answer, $x^4-2x^3+3x^2-2x+1$.

15. x^4 $+x^3$ $+x^2$.

$-x^3$ $-x^2$ $-x$.

x^2 x 1. Answer x^4+x^2+1.

16. (i) $x^4-2x^2y^2+y^4$.

(ii) $4x^2y^2$.

(iii) $x^4+2x^2y^2+y^4$.

Triangle will be right-angled (see Chapter 6).

Take $x=2$, $y=1$.

A CATALOG OF SELECTED
DOVER BOOKS
IN ALL FIELDS OF INTEREST

A CATALOG OF SELECTED DOVER
BOOKS IN ALL FIELDS OF INTEREST

CONCERNING THE SPIRITUAL IN ART, Wassily Kandinsky. Pioneering work by father of abstract art. Thoughts on color theory, nature of art. Analysis of earlier masters. 12 illustrations. 80pp. of text. 5⅜ x 8½. 23411-8

ANIMALS: 1,419 Copyright-Free Illustrations of Mammals, Birds, Fish, Insects, etc., Jim Harter (ed.). Clear wood engravings present, in extremely lifelike poses, over 1,000 species of animals. One of the most extensive pictorial sourcebooks of its kind. Captions. Index. 284pp. 9 x 12. 23766-4

CELTIC ART: The Methods of Construction, George Bain. Simple geometric techniques for making Celtic interlacements, spirals, Kells-type initials, animals, humans, etc. Over 500 illustrations. 160pp. 9 x 12. (Available in U.S. only.) 22923-8

AN ATLAS OF ANATOMY FOR ARTISTS, Fritz Schider. Most thorough reference work on art anatomy in the world. Hundreds of illustrations, including selections from works by Vesalius, Leonardo, Goya, Ingres, Michelangelo, others. 593 illustrations. 192pp. 7⅛ x 10¼. 20241-0

CELTIC HAND STROKE-BY-STROKE (Irish Half-Uncial from "The Book of Kells"): An Arthur Baker Calligraphy Manual, Arthur Baker. Complete guide to creating each letter of the alphabet in distinctive Celtic manner. Covers hand position, strokes, pens, inks, paper, more. Illustrated. 48pp. 8¼ x 11. 24336-2

EASY ORIGAMI, John Montroll. Charming collection of 32 projects (hat, cup, pelican, piano, swan, many more) specially designed for the novice origami hobbyist. Clearly illustrated easy-to-follow instructions insure that even beginning papercrafters will achieve successful results. 48pp. 8¼ x 11. 27298-2

THE COMPLETE BOOK OF BIRDHOUSE CONSTRUCTION FOR WOOD-WORKERS, Scott D. Campbell. Detailed instructions, illustrations, tables. Also data on bird habitat and instinct patterns. Bibliography. 3 tables. 63 illustrations in 15 figures. 48pp. 5¼ x 8½. 24407-5

BLOOMINGDALE'S ILLUSTRATED 1886 CATALOG: Fashions, Dry Goods and Housewares, Bloomingdale Brothers. Famed merchants' extremely rare catalog depicting about 1,700 products: clothing, housewares, firearms, dry goods, jewelry, more. Invaluable for dating, identifying vintage items. Also, copyright-free graphics for artists, designers. Co-published with Henry Ford Museum & Greenfield Village. 160pp. 8¼ x 11. 25780-0

HISTORIC COSTUME IN PICTURES, Braun & Schneider. Over 1,450 costumed figures in clearly detailed engravings–from dawn of civilization to end of 19th century. Captions. Many folk costumes. 256pp. 8⅜ x 11¼. 23150-X

STICKLEY CRAFTSMAN FURNITURE CATALOGS, Gustav Stickley and L. & J. G. Stickley. Beautiful, functional furniture in two authentic catalogs from 1910. 594 illustrations, including 277 photos, show settles, rockers, armchairs, reclining chairs, bookcases, desks, tables. 183pp. 6½ x 9¼. 23838-5

AMERICAN LOCOMOTIVES IN HISTORIC PHOTOGRAPHS: 1858 to 1949, Ron Ziel (ed.). A rare collection of 126 meticulously detailed official photographs, called "builder portraits," of American locomotives that majestically chronicle the rise of steam locomotive power in America. Introduction. Detailed captions. xi+ 129pp. 9 x 12. 27393-8

AMERICA'S LIGHTHOUSES: An Illustrated History, Francis Ross Holland, Jr. Delightfully written, profusely illustrated fact-filled survey of over 200 American lighthouses since 1716. History, anecdotes, technological advances, more. 240pp. 8 x 10¾. 25576-X

TOWARDS A NEW ARCHITECTURE, Le Corbusier. Pioneering manifesto by founder of "International School." Technical and aesthetic theories, views of industry, economics, relation of form to function, "mass-production split" and much more. Profusely illustrated. 320pp. 6⅛ x 9¼. (Available in U.S. only.) 25023-7

HOW THE OTHER HALF LIVES, Jacob Riis. Famous journalistic record, exposing poverty and degradation of New York slums around 1900, by major social reformer. 100 striking and influential photographs. 233pp. 10 x 7⅞. 22012-5

FRUIT KEY AND TWIG KEY TO TREES AND SHRUBS, William M. Harlow. One of the handiest and most widely used identification aids. Fruit key covers 120 deciduous and evergreen species; twig key 160 deciduous species. Easily used. Over 300 photographs. 126pp. 5⅜ x 8½. 20511-8

COMMON BIRD SONGS, Dr. Donald J. Borror. Songs of 60 most common U.S. birds: robins, sparrows, cardinals, bluejays, finches, more—arranged in order of increasing complexity. Up to 9 variations of songs of each species.
Cassette and manual 99911-4

ORCHIDS AS HOUSE PLANTS, Rebecca Tyson Northen. Grow cattleyas and many other kinds of orchids—in a window, in a case, or under artificial light. 63 illustrations. 148pp. 5⅜ x 8½. 23261-1

MONSTER MAZES, Dave Phillips. Masterful mazes at four levels of difficulty. Avoid deadly perils and evil creatures to find magical treasures. Solutions for all 32 exciting illustrated puzzles. 48pp. 8¼ x 11. 26005-4

MOZART'S DON GIOVANNI (DOVER OPERA LIBRETTO SERIES), Wolfgang Amadeus Mozart. Introduced and translated by Ellen H. Bleiler. Standard Italian libretto, with complete English translation. Convenient and thoroughly portable—an ideal companion for reading along with a recording or the performance itself. Introduction. List of characters. Plot summary. 121pp. 5¼ x 8½. 24944-1

TECHNICAL MANUAL AND DICTIONARY OF CLASSICAL BALLET, Gail Grant. Defines, explains, comments on steps, movements, poses and concepts. 15-page pictorial section. Basic book for student, viewer. 127pp. 5⅜ x 8½. 21843-0

CATALOG OF DOVER BOOKS

THE CLARINET AND CLARINET PLAYING, David Pino. Lively, comprehensive work features suggestions about technique, musicianship, and musical interpretation, as well as guidelines for teaching, making your own reeds, and preparing for public performance. Includes an intriguing look at clarinet history. "A godsend," *The Clarinet,* Journal of the International Clarinet Society. Appendixes. 7 illus. 320pp. 5⅜ x 8½. 40270-3

HOLLYWOOD GLAMOR PORTRAITS, John Kobal (ed.). 145 photos from 1926-49. Harlow, Gable, Bogart, Bacall; 94 stars in all. Full background on photographers, technical aspects. 160pp. 8⅜ x 11¼. 23352-9

THE ANNOTATED CASEY AT THE BAT: A Collection of Ballads about the Mighty Casey/Third, Revised Edition, Martin Gardner (ed.). Amusing sequels and parodies of one of America's best-loved poems: Casey's Revenge, Why Casey Whiffed, Casey's Sister at the Bat, others. 256pp. 5⅜ x 8½. 28598-7

THE RAVEN AND OTHER FAVORITE POEMS, Edgar Allan Poe. Over 40 of the author's most memorable poems: "The Bells," "Ulalume," "Israfel," "To Helen," "The Conqueror Worm," "Eldorado," "Annabel Lee," many more. Alphabetic lists of titles and first lines. 64pp. 5³⁄₁₆ x 8¼. 26685-0

PERSONAL MEMOIRS OF U. S. GRANT, Ulysses Simpson Grant. Intelligent, deeply moving firsthand account of Civil War campaigns, considered by many the finest military memoirs ever written. Includes letters, historic photographs, maps and more. 528pp. 6⅛ x 9¼. 28587-1

ANCIENT EGYPTIAN MATERIALS AND INDUSTRIES, A. Lucas and J. Harris. Fascinating, comprehensive, thoroughly documented text describes this ancient civilization's vast resources and the processes that incorporated them in daily life, including the use of animal products, building materials, cosmetics, perfumes and incense, fibers, glazed ware, glass and its manufacture, materials used in the mummification process, and much more. 544pp. 6⅛ x 9¼. (Available in U.S. only.) 40446-3

RUSSIAN STORIES/RUSSKIE RASSKAZY: A Dual-Language Book, edited by Gleb Struve. Twelve tales by such masters as Chekhov, Tolstoy, Dostoevsky, Pushkin, others. Excellent word-for-word English translations on facing pages, plus teaching and study aids, Russian/English vocabulary, biographical/critical introductions, more. 416pp. 5⅜ x 8½. 26244-8

PHILADELPHIA THEN AND NOW: 60 Sites Photographed in the Past and Present, Kenneth Finkel and Susan Oyama. Rare photographs of City Hall, Logan Square, Independence Hall, Betsy Ross House, other landmarks juxtaposed with contemporary views. Captures changing face of historic city. Introduction. Captions. 128pp. 8¼ x 11. 25790-8

AIA ARCHITECTURAL GUIDE TO NASSAU AND SUFFOLK COUNTIES, LONG ISLAND, The American Institute of Architects, Long Island Chapter, and the Society for the Preservation of Long Island Antiquities. Comprehensive, well-researched and generously illustrated volume brings to life over three centuries of Long Island's great architectural heritage. More than 240 photographs with authoritative, extensively detailed captions. 176pp. 8¼ x 11. 26946-9

NORTH AMERICAN INDIAN LIFE: Customs and Traditions of 23 Tribes, Elsie Clews Parsons (ed.). 27 fictionalized essays by noted anthropologists examine religion, customs, government, additional facets of life among the Winnebago, Crow, Zuni, Eskimo, other tribes. 480pp. 6⅛ x 9¼. 27377-6

FRANK LLOYD WRIGHT'S DANA HOUSE, Donald Hoffmann. Pictorial essay of residential masterpiece with over 160 interior and exterior photos, plans, elevations, sketches and studies. 128pp. 9¼ x 10¾. 29120-0

THE MALE AND FEMALE FIGURE IN MOTION: 60 Classic Photographic Sequences, Eadweard Muybridge. 60 true-action photographs of men and women walking, running, climbing, bending, turning, etc., reproduced from rare 19th-century masterpiece. vi + 121pp. 9 x 12. 24745-7

1001 QUESTIONS ANSWERED ABOUT THE SEASHORE, N. J. Berrill and Jacquelyn Berrill. Queries answered about dolphins, sea snails, sponges, starfish, fishes, shore birds, many others. Covers appearance, breeding, growth, feeding, much more. 305pp. 5¼ x 8¼. 23366-9

ATTRACTING BIRDS TO YOUR YARD, William J. Weber. Easy-to-follow guide offers advice on how to attract the greatest diversity of birds: birdhouses, feeders, water and waterers, much more. 96pp. 5³⁄₁₆ x 8¼. 28927-3

MEDICINAL AND OTHER USES OF NORTH AMERICAN PLANTS: A Historical Survey with Special Reference to the Eastern Indian Tribes, Charlotte Erichsen-Brown. Chronological historical citations document 500 years of usage of plants, trees, shrubs native to eastern Canada, northeastern U.S. Also complete identifying information. 343 illustrations. 544pp. 6½ x 9¼. 25951-X

STORYBOOK MAZES, Dave Phillips. 23 stories and mazes on two-page spreads: Wizard of Oz, Treasure Island, Robin Hood, etc. Solutions. 64pp. 8¼ x 11. 23628-5

AMERICAN NEGRO SONGS: 230 Folk Songs and Spirituals, Religious and Secular, John W. Work. This authoritative study traces the African influences of songs sung and played by black Americans at work, in church, and as entertainment. The author discusses the lyric significance of such songs as "Swing Low, Sweet Chariot," "John Henry," and others and offers the words and music for 230 songs. Bibliography. Index of Song Titles. 272pp. 6½ x 9¼. 40271-1

MOVIE-STAR PORTRAITS OF THE FORTIES, John Kobal (ed.). 163 glamor, studio photos of 106 stars of the 1940s: Rita Hayworth, Ava Gardner, Marlon Brando, Clark Gable, many more. 176pp. 8⅜ x 11¼. 23546-7

BENCHLEY LOST AND FOUND, Robert Benchley. Finest humor from early 30s, about pet peeves, child psychologists, post office and others. Mostly unavailable elsewhere. 73 illustrations by Peter Arno and others. 183pp. 5⅜ x 8½. 22410-4

YEKL and THE IMPORTED BRIDEGROOM AND OTHER STORIES OF YIDDISH NEW YORK, Abraham Cahan. Film Hester Street based on *Yekl* (1896). Novel, other stories among first about Jewish immigrants on N.Y.'s East Side. 240pp. 5⅜ x 8½. 22427-9

SELECTED POEMS, Walt Whitman. Generous sampling from *Leaves of Grass*. Twenty-four poems include "I Hear America Singing," "Song of the Open Road," "I Sing the Body Electric," "When Lilacs Last in the Dooryard Bloom'd," "O Captain! My Captain!"–all reprinted from an authoritative edition. Lists of titles and first lines. 128pp. 5³⁄₁₆ x 8¼. 26878-0

THE BEST TALES OF HOFFMANN, E. T. A. Hoffmann. 10 of Hoffmann's most important stories: "Nutcracker and the King of Mice," "The Golden Flowerpot," etc. 458pp. 5⅜ x 8½. 21793-0

FROM FETISH TO GOD IN ANCIENT EGYPT, E. A. Wallis Budge. Rich detailed survey of Egyptian conception of "God" and gods, magic, cult of animals, Osiris, more. Also, superb English translations of hymns and legends. 240 illustrations. 545pp. 5⅜ x 8½. 25803-3

FRENCH STORIES/CONTES FRANÇAIS: A Dual-Language Book, Wallace Fowlie. Ten stories by French masters, Voltaire to Camus: "Micromegas" by Voltaire; "The Atheist's Mass" by Balzac; "Minuet" by de Maupassant; "The Guest" by Camus, six more. Excellent English translations on facing pages. Also French-English vocabulary list, exercises, more. 352pp. 5⅜ x 8½. 26443-2

CHICAGO AT THE TURN OF THE CENTURY IN PHOTOGRAPHS: 122 Historic Views from the Collections of the Chicago Historical Society, Larry A. Viskochil. Rare large-format prints offer detailed views of City Hall, State Street, the Loop, Hull House, Union Station, many other landmarks, circa 1904-1913. Introduction. Captions. Maps. 144pp. 9⅜ x 12¼. 24656-6

OLD BROOKLYN IN EARLY PHOTOGRAPHS, 1865-1929, William Lee Younger. Luna Park, Gravesend race track, construction of Grand Army Plaza, moving of Hotel Brighton, etc. 157 previously unpublished photographs. 165pp. 8⅞ x 11¾. 23587-4

THE MYTHS OF THE NORTH AMERICAN INDIANS, Lewis Spence. Rich anthology of the myths and legends of the Algonquins, Iroquois, Pawnees and Sioux, prefaced by an extensive historical and ethnological commentary. 36 illustrations. 480pp. 5⅜ x 8½. 25967-6

AN ENCYCLOPEDIA OF BATTLES: Accounts of Over 1,560 Battles from 1479 B.C. to the Present, David Eggenberger. Essential details of every major battle in recorded history from the first battle of Megiddo in 1479 B.C. to Grenada in 1984. List of Battle Maps. New Appendix covering the years 1967-1984. Index. 99 illustrations. 544pp. 6½ x 9¼. 24913-1

SAILING ALONE AROUND THE WORLD, Captain Joshua Slocum. First man to sail around the world, alone, in small boat. One of great feats of seamanship told in delightful manner. 67 illustrations. 294pp. 5⅜ x 8½. 20326-3

ANARCHISM AND OTHER ESSAYS, Emma Goldman. Powerful, penetrating, prophetic essays on direct action, role of minorities, prison reform, puritan hypocrisy, violence, etc. 271pp. 5⅜ x 8½. 22484-8

MYTHS OF THE HINDUS AND BUDDHISTS, Ananda K. Coomaraswamy and Sister Nivedita. Great stories of the epics; deeds of Krishna, Shiva, taken from puranas, Vedas, folk tales; etc. 32 illustrations. 400pp. 5⅜ x 8½. 21759-0

THE TRAUMA OF BIRTH, Otto Rank. Rank's controversial thesis that anxiety neurosis is caused by profound psychological trauma which occurs at birth. 256pp. 5⅜ x 8½. 27974-X

A THEOLOGICO-POLITICAL TREATISE, Benedict Spinoza. Also contains unfinished Political Treatise. Great classic on religious liberty, theory of government on common consent. R. Elwes translation. Total of 421pp. 5⅜ x 8½. 20249-6

CATALOG OF DOVER BOOKS

MY BONDAGE AND MY FREEDOM, Frederick Douglass. Born a slave, Douglass became outspoken force in antislavery movement. The best of Douglass' autobiographies. Graphic description of slave life. 464pp. 5⅜ x 8½. 22457-0

FOLLOWING THE EQUATOR: A Journey Around the World, Mark Twain. Fascinating humorous account of 1897 voyage to Hawaii, Australia, India, New Zealand, etc. Ironic, bemused reports on peoples, customs, climate, flora and fauna, politics, much more. 197 illustrations. 720pp. 5⅜ x 8½. 26113-1

THE PEOPLE CALLED SHAKERS, Edward D. Andrews. Definitive study of Shakers: origins, beliefs, practices, dances, social organization, furniture and crafts, etc. 33 illustrations. 351pp. 5⅜ x 8½. 21081-2

THE MYTHS OF GREECE AND ROME, H. A. Guerber. A classic of mythology, generously illustrated, long prized for its simple, graphic, accurate retelling of the principal myths of Greece and Rome, and for its commentary on their origins and significance. With 64 illustrations by Michelangelo, Raphael, Titian, Rubens, Canova, Bernini and others. 480pp. 5⅜ x 8½. 27584-1

PSYCHOLOGY OF MUSIC, Carl E. Seashore. Classic work discusses music as a medium from psychological viewpoint. Clear treatment of physical acoustics, auditory apparatus, sound perception, development of musical skills, nature of musical feeling, host of other topics. 88 figures. 408pp. 5⅜ x 8½. 21851-1

THE PHILOSOPHY OF HISTORY, Georg W. Hegel. Great classic of Western thought develops concept that history is not chance but rational process, the evolution of freedom. 457pp. 5⅜ x 8½. 20112-0

THE BOOK OF TEA, Kakuzo Okakura. Minor classic of the Orient: entertaining, charming explanation, interpretation of traditional Japanese culture in terms of tea ceremony. 94pp. 5⅜ x 8½. 20070-1

LIFE IN ANCIENT EGYPT, Adolf Erman. Fullest, most thorough, detailed older account with much not in more recent books, domestic life, religion, magic, medicine, commerce, much more. Many illustrations reproduce tomb paintings, carvings, hieroglyphs, etc. 597pp. 5⅜ x 8½. 22632-8

SUNDIALS, Their Theory and Construction, Albert Waugh. Far and away the best, most thorough coverage of ideas, mathematics concerned, types, construction, adjusting anywhere. Simple, nontechnical treatment allows even children to build several of these dials. Over 100 illustrations. 230pp. 5⅜ x 8½. 22947-5

THEORETICAL HYDRODYNAMICS, L. M. Milne-Thomson. Classic exposition of the mathematical theory of fluid motion, applicable to both hydrodynamics and aerodynamics. Over 600 exercises. 768pp. 6⅛ x 9¼. 68970-0

SONGS OF EXPERIENCE: Facsimile Reproduction with 26 Plates in Full Color, William Blake. 26 full-color plates from a rare 1826 edition. Includes "The Tyger," "London," "Holy Thursday," and other poems. Printed text of poems. 48pp. 5¼ x 7. 24636-1

OLD-TIME VIGNETTES IN FULL COLOR, Carol Belanger Grafton (ed.). Over 390 charming, often sentimental illustrations, selected from archives of Victorian graphics—pretty women posing, children playing, food, flowers, kittens and puppies, smiling cherubs, birds and butterflies, much more. All copyright-free. 48pp. 9¼ x 12¼. 27269-9

PERSPECTIVE FOR ARTISTS, Rex Vicat Cole. Depth, perspective of sky and sea, shadows, much more, not usually covered. 391 diagrams, 81 reproductions of drawings and paintings. 279pp. 5⅜ x 8½. 22487-2

DRAWING THE LIVING FIGURE, Joseph Sheppard. Innovative approach to artistic anatomy focuses on specifics of surface anatomy, rather than muscles and bones. Over 170 drawings of live models in front, back and side views, and in widely varying poses. Accompanying diagrams. 177 illustrations. Introduction. Index. 144pp. 8⅜ x11¼. 26723-7

GOTHIC AND OLD ENGLISH ALPHABETS: 100 Complete Fonts, Dan X. Solo. Add power, elegance to posters, signs, other graphics with 100 stunning copyright-free alphabets: Blackstone, Dolbey, Germania, 97 more—including many lower-case, numerals, punctuation marks. 104pp. 8⅛ x 11. 24695-7

HOW TO DO BEADWORK, Mary White. Fundamental book on craft from simple projects to five-bead chains and woven works. 106 illustrations. 142pp. 5⅜ x 8. 20697-1

THE BOOK OF WOOD CARVING, Charles Marshall Sayers. Finest book for beginners discusses fundamentals and offers 34 designs. "Absolutely first rate . . . well thought out and well executed."—E. J. Tangerman. 118pp. 7¾ x 10⅝. 23654-4

ILLUSTRATED CATALOG OF CIVIL WAR MILITARY GOODS: Union Army Weapons, Insignia, Uniform Accessories, and Other Equipment, Schuyler, Hartley, and Graham. Rare, profusely illustrated 1846 catalog includes Union Army uniform and dress regulations, arms and ammunition, coats, insignia, flags, swords, rifles, etc. 226 illustrations. 160pp. 9 x 12. 24939-5

WOMEN'S FASHIONS OF THE EARLY 1900s: An Unabridged Republication of "New York Fashions, 1909," National Cloak & Suit Co. Rare catalog of mail-order fashions documents women's and children's clothing styles shortly after the turn of the century. Captions offer full descriptions, prices. Invaluable resource for fashion, costume historians. Approximately 725 illustrations. 128pp. 8⅜ x 11¼. 27276-1

THE 1912 AND 1915 GUSTAV STICKLEY FURNITURE CATALOGS, Gustav Stickley. With over 200 detailed illustrations and descriptions, these two catalogs are essential reading and reference materials and identification guides for Stickley furniture. Captions cite materials, dimensions and prices. 112pp. 6½ x 9¼. 26676-1

EARLY AMERICAN LOCOMOTIVES, John H. White, Jr. Finest locomotive engravings from early 19th century: historical (1804–74), main-line (after 1870), special, foreign, etc. 147 plates. 142pp. 11⅜ x 8¼. 22772-3

THE TALL SHIPS OF TODAY IN PHOTOGRAPHS, Frank O. Braynard. Lavishly illustrated tribute to nearly 100 majestic contemporary sailing vessels: Amerigo Vespucci, Clearwater, Constitution, Eagle, Mayflower, Sea Cloud, Victory, many more. Authoritative captions provide statistics, background on each ship. 190 black-and-white photographs and illustrations. Introduction. 128pp. 8⅞ x 11¾. 27163-3

LITTLE BOOK OF EARLY AMERICAN CRAFTS AND TRADES, Peter Stockham (ed.). 1807 children's book explains crafts and trades: baker, hatter, cooper, potter, and many others. 23 copperplate illustrations. 140pp. $4^5/_8$ x 6.   23336-7

VICTORIAN FASHIONS AND COSTUMES FROM HARPER'S BAZAR, 1867–1898, Stella Blum (ed.). Day costumes, evening wear, sports clothes, shoes, hats, other accessories in over 1,000 detailed engravings. 320pp. $9\frac{3}{8}$ x $12\frac{1}{4}$.   22990-4

GUSTAV STICKLEY, THE CRAFTSMAN, Mary Ann Smith. Superb study surveys broad scope of Stickley's achievement, especially in architecture. Design philosophy, rise and fall of the Craftsman empire, descriptions and floor plans for many Craftsman houses, more. 86 black-and-white halftones. 31 line illustrations. Introduction 208pp. $6\frac{1}{2}$ x $9\frac{1}{4}$.   27210-9

THE LONG ISLAND RAIL ROAD IN EARLY PHOTOGRAPHS, Ron Ziel. Over 220 rare photos, informative text document origin (1844) and development of rail service on Long Island. Vintage views of early trains, locomotives, stations, passengers, crews, much more. Captions. $8\frac{7}{8}$ x $11\frac{3}{4}$.   26301-0

VOYAGE OF THE LIBERDADE, Joshua Slocum. Great 19th-century mariner's thrilling, first-hand account of the wreck of his ship off South America, the 35-foot boat he built from the wreckage, and its remarkable voyage home. 128pp. $5\frac{3}{8}$ x $8\frac{1}{2}$.   40022-0

TEN BOOKS ON ARCHITECTURE, Vitruvius. The most important book ever written on architecture. Early Roman aesthetics, technology, classical orders, site selection, all other aspects. Morgan translation. 331pp. $5\frac{3}{8}$ x $8\frac{1}{2}$.   20645-9

THE HUMAN FIGURE IN MOTION, Eadweard Muybridge. More than 4,500 stopped-action photos, in action series, showing undraped men, women, children jumping, lying down, throwing, sitting, wrestling, carrying, etc. 390pp. $7\frac{7}{8}$ x $10\frac{5}{8}$.   20204-6 Clothbd.

TREES OF THE EASTERN AND CENTRAL UNITED STATES AND CANADA, William M. Harlow. Best one-volume guide to 140 trees. Full descriptions, woodlore, range, etc. Over 600 illustrations. Handy size. 288pp. $4\frac{1}{2}$ x $6\frac{3}{4}$.   20395-6

SONGS OF WESTERN BIRDS, Dr. Donald J. Borror. Complete song and call repertoire of 60 western species, including flycatchers, juncoes, cactus wrens, many more–includes fully illustrated booklet.   Cassette and manual 99913-0

GROWING AND USING HERBS AND SPICES, Milo Miloradovich. Versatile handbook provides all the information needed for cultivation and use of all the herbs and spices available in North America. 4 illustrations. Index. Glossary. 236pp. $5\frac{3}{8}$ x $8\frac{1}{2}$.   25058-X

BIG BOOK OF MAZES AND LABYRINTHS, Walter Shepherd. 50 mazes and labyrinths in all–classical, solid, ripple, and more–in one great volume. Perfect inexpensive puzzler for clever youngsters. Full solutions. 112pp. $8\frac{1}{4}$ x 11.   22951-3

PIANO TUNING, J. Cree Fischer. Clearest, best book for beginner, amateur. Simple repairs, raising dropped notes, tuning by easy method of flattened fifths. No previous skills needed. 4 illustrations. 201pp. 5⅜ x 8½. 23267-0

HINTS TO SINGERS, Lillian Nordica. Selecting the right teacher, developing confidence, overcoming stage fright, and many other important skills receive thoughtful discussion in this indispensible guide, written by a world-famous diva of four decades' experience. 96pp. 5⅜ x 8½. 40094-8

THE COMPLETE NONSENSE OF EDWARD LEAR, Edward Lear. All nonsense limericks, zany alphabets, Owl and Pussycat, songs, nonsense botany, etc., illustrated by Lear. Total of 320pp. 5⅜ x 8½. (Available in U.S. only.) 20167-8

VICTORIAN PARLOUR POETRY: An Annotated Anthology, Michael R. Turner. 117 gems by Longfellow, Tennyson, Browning, many lesser-known poets. "The Village Blacksmith," "Curfew Must Not Ring Tonight," "Only a Baby Small," dozens more, often difficult to find elsewhere. Index of poets, titles, first lines. xxiii + 325pp. 5⅜ x 8¼. 27044-0

DUBLINERS, James Joyce. Fifteen stories offer vivid, tightly focused observations of the lives of Dublin's poorer classes. At least one, "The Dead," is considered a masterpiece. Reprinted complete and unabridged from standard edition. 160pp. 5³⁄₁₆ x 8¼. 26870-5

GREAT WEIRD TALES: 14 Stories by Lovecraft, Blackwood, Machen and Others, S. T. Joshi (ed.). 14 spellbinding tales, including "The Sin Eater," by Fiona McLeod, "The Eye Above the Mantel," by Frank Belknap Long, as well as renowned works by R. H. Barlow, Lord Dunsany, Arthur Machen, W. C. Morrow and eight other masters of the genre. 256pp. 5⅜ x 8½. (Available in U.S. only.) 40436-6

THE BOOK OF THE SACRED MAGIC OF ABRAMELIN THE MAGE, translated by S. MacGregor Mathers. Medieval manuscript of ceremonial magic. Basic document in Aleister Crowley, Golden Dawn groups. 268pp. 5⅜ x 8½. 23211-5

NEW RUSSIAN-ENGLISH AND ENGLISH-RUSSIAN DICTIONARY, M. A. O'Brien. This is a remarkably handy Russian dictionary, containing a surprising amount of information, including over 70,000 entries. 366pp. 4½ x 6⅛. 20208-9

HISTORIC HOMES OF THE AMERICAN PRESIDENTS, Second, Revised Edition, Irvin Haas. A traveler's guide to American Presidential homes, most open to the public, depicting and describing homes occupied by every American President from George Washington to George Bush. With visiting hours, admission charges, travel routes. 175 photographs. Index. 160pp. 8¼ x 11. 26751-2

NEW YORK IN THE FORTIES, Andreas Feininger. 162 brilliant photographs by the well-known photographer, formerly with *Life* magazine. Commuters, shoppers, Times Square at night, much else from city at its peak. Captions by John von Hartz. 181pp. 9¼ x 10¾. 23585-8

INDIAN SIGN LANGUAGE, William Tomkins. Over 525 signs developed by Sioux and other tribes. Written instructions and diagrams. Also 290 pictographs. 111pp. 6⅛ x 9¼. 22029-X

CATALOG OF DOVER BOOKS

ANATOMY: A Complete Guide for Artists, Joseph Sheppard. A master of figure drawing shows artists how to render human anatomy convincingly. Over 460 illustrations. 224pp. 8⅜ x 11¼. 27279-6

MEDIEVAL CALLIGRAPHY: Its History and Technique, Marc Drogin. Spirited history, comprehensive instruction manual covers 13 styles (ca. 4th century through 15th). Excellent photographs; directions for duplicating medieval techniques with modern tools. 224pp. 8⅜ x 11¼. 26142-5

DRIED FLOWERS: How to Prepare Them, Sarah Whitlock and Martha Rankin. Complete instructions on how to use silica gel, meal and borax, perlite aggregate, sand and borax, glycerine and water to create attractive permanent flower arrangements. 12 illustrations. 32pp. 5⅜ x 8½. 21802-3

EASY-TO-MAKE BIRD FEEDERS FOR WOODWORKERS, Scott D. Campbell. Detailed, simple-to-use guide for designing, constructing, caring for and using feeders. Text, illustrations for 12 classic and contemporary designs. 96pp. 5⅜ x 8½. 25847-5

SCOTTISH WONDER TALES FROM MYTH AND LEGEND, Donald A. Mackenzie. 16 lively tales tell of giants rumbling down mountainsides, of a magic wand that turns stone pillars into warriors, of gods and goddesses, evil hags, powerful forces and more. 240pp. 5⅜ x 8½. 29677-6

THE HISTORY OF UNDERCLOTHES, C. Willett Cunnington and Phyllis Cunnington. Fascinating, well-documented survey covering six centuries of English undergarments, enhanced with over 100 illustrations: 12th-century laced-up bodice, footed long drawers (1795), 19th-century bustles, 19th-century corsets for men, Victorian "bust improvers," much more. 272pp. 5⅝ x 8¼. 27124-2

ARTS AND CRAFTS FURNITURE: The Complete Brooks Catalog of 1912, Brooks Manufacturing Co. Photos and detailed descriptions of more than 150 now very collectible furniture designs from the Arts and Crafts movement depict davenports, settees, buffets, desks, tables, chairs, bedsteads, dressers and more, all built of solid, quarter-sawed oak. Invaluable for students and enthusiasts of antiques, Americana and the decorative arts. 80pp. 6½ x 9¼. 27471-3

WILBUR AND ORVILLE: A Biography of the Wright Brothers, Fred Howard. Definitive, crisply written study tells the full story of the brothers' lives and work. A vividly written biography, unparalleled in scope and color, that also captures the spirit of an extraordinary era. 560pp. 6⅛ x 9¼. 40297-5

THE ARTS OF THE SAILOR: Knotting, Splicing and Ropework, Hervey Garrett Smith. Indispensable shipboard reference covers tools, basic knots and useful hitches; handsewing and canvas work, more. Over 100 illustrations. Delightful reading for sea lovers. 256pp. 5⅜ x 8½. 26440-8

FRANK LLOYD WRIGHT'S FALLINGWATER: The House and Its History, Second, Revised Edition, Donald Hoffmann. A total revision–both in text and illustrations–of the standard document on Fallingwater, the boldest, most personal architectural statement of Wright's mature years, updated with valuable new material from the recently opened Frank Lloyd Wright Archives. "Fascinating"–*The New York Times*. 116 illustrations. 128pp. 9¼ x 10¾. 27430-6

PHOTOGRAPHIC SKETCHBOOK OF THE CIVIL WAR, Alexander Gardner. 100 photos taken on field during the Civil War. Famous shots of Manassas Harper's Ferry, Lincoln, Richmond, slave pens, etc. 244pp. 10¾ x 8¼. 22731-6

FIVE ACRES AND INDEPENDENCE, Maurice G. Kains. Great back-to-the-land classic explains basics of self-sufficient farming. The one book to get. 95 illustrations. 397pp. 5⅜ x 8½. 20974-1

SONGS OF EASTERN BIRDS, Dr. Donald J. Borror. Songs and calls of 60 species most common to eastern U.S.: warblers, woodpeckers, flycatchers, thrushes, larks, many more in high-quality recording. Cassette and manual 99912-2

A MODERN HERBAL, Margaret Grieve. Much the fullest, most exact, most useful compilation of herbal material. Gigantic alphabetical encyclopedia, from aconite to zedoary, gives botanical information, medical properties, folklore, economic uses, much else. Indispensable to serious reader. 161 illustrations. 888pp. 6½ x 9¼. 2-vol. set. (Available in U.S. only.) Vol. I: 22798-7
Vol. II: 22799-5

HIDDEN TREASURE MAZE BOOK, Dave Phillips. Solve 34 challenging mazes accompanied by heroic tales of adventure. Evil dragons, people-eating plants, blood-thirsty giants, many more dangerous adversaries lurk at every twist and turn. 34 mazes, stories, solutions. 48pp. 8¼ x 11. 24566-7

LETTERS OF W. A. MOZART, Wolfgang A. Mozart. Remarkable letters show bawdy wit, humor, imagination, musical insights, contemporary musical world; includes some letters from Leopold Mozart. 276pp. 5⅜ x 8½. 22859-2

BASIC PRINCIPLES OF CLASSICAL BALLET, Agrippina Vaganova. Great Russian theoretician, teacher explains methods for teaching classical ballet. 118 illustrations. 175pp. 5⅜ x 8½. 22036-2

THE JUMPING FROG, Mark Twain. Revenge edition. The original story of The Celebrated Jumping Frog of Calaveras County, a hapless French translation, and Twain's hilarious "retranslation" from the French. 12 illustrations. 66pp. 5⅜ x 8½. 22686-7

BEST REMEMBERED POEMS, Martin Gardner (ed.). The 126 poems in this superb collection of 19th- and 20th-century British and American verse range from Shelley's "To a Skylark" to the impassioned "Renascence" of Edna St. Vincent Millay and to Edward Lear's whimsical "The Owl and the Pussycat." 224pp. 5⅜ x 8½. 27165-X

COMPLETE SONNETS, William Shakespeare. Over 150 exquisite poems deal with love, friendship, the tyranny of time, beauty's evanescence, death and other themes in language of remarkable power, precision and beauty. Glossary of archaic terms. 80pp. 5³⁄₁₆ x 8¼. 26686-9

THE BATTLES THAT CHANGED HISTORY, Fletcher Pratt. Eminent historian profiles 16 crucial conflicts, ancient to modern, that changed the course of civilization. 352pp. 5⅜ x 8½. 41129-X

THE WIT AND HUMOR OF OSCAR WILDE, Alvin Redman (ed.). More than 1,000 ripostes, paradoxes, wisecracks: Work is the curse of the drinking classes; I can resist everything except temptation; etc. 258pp. 5⅜ x 8½. 20602-5

SHAKESPEARE LEXICON AND QUOTATION DICTIONARY, Alexander Schmidt. Full definitions, locations, shades of meaning in every word in plays and poems. More than 50,000 exact quotations. 1,485pp. 6½ x 9¼. 2-vol. set.
Vol. 1: 22726-X
Vol. 2: 22727-8

SELECTED POEMS, Emily Dickinson. Over 100 best-known, best-loved poems by one of America's foremost poets, reprinted from authoritative early editions. No comparable edition at this price. Index of first lines. 64pp. 5³⁄₁₆ x 8¼. 26466-1

THE INSIDIOUS DR. FU-MANCHU, Sax Rohmer. The first of the popular mystery series introduces a pair of English detectives to their archnemesis, the diabolical Dr. Fu-Manchu. Flavorful atmosphere, fast-paced action, and colorful characters enliven this classic of the genre. 208pp. 5³⁄₁₆ x 8¼. 29898-1

THE MALLEUS MALEFICARUM OF KRAMER AND SPRENGER, translated by Montague Summers. Full text of most important witchhunter's "bible," used by both Catholics and Protestants. 278pp. 6⅝ x 10. 22802-9

SPANISH STORIES/CUENTOS ESPAÑOLES: A Dual-Language Book, Angel Flores (ed.). Unique format offers 13 great stories in Spanish by Cervantes, Borges, others. Faithful English translations on facing pages. 352pp. 5⅜ x 8½. 25399-6

GARDEN CITY, LONG ISLAND, IN EARLY PHOTOGRAPHS, 1869–1919, Mildred H. Smith. Handsome treasury of 118 vintage pictures, accompanied by carefully researched captions, document the Garden City Hotel fire (1899), the Vanderbilt Cup Race (1908), the first airmail flight departing from the Nassau Boulevard Aerodrome (1911), and much more. 96pp. 8⅞ x 11¾. 40669-5

OLD QUEENS, N.Y., IN EARLY PHOTOGRAPHS, Vincent F. Seyfried and William Asadorian. Over 160 rare photographs of Maspeth, Jamaica, Jackson Heights, and other areas. Vintage views of DeWitt Clinton mansion, 1939 World's Fair and more. Captions. 192pp. 8⅞ x 11. 26358-4

CAPTURED BY THE INDIANS: 15 Firsthand Accounts, 1750-1870, Frederick Drimmer. Astounding true historical accounts of grisly torture, bloody conflicts, relentless pursuits, miraculous escapes and more, by people who lived to tell the tale. 384pp. 5⅜ x 8½. 24901-8

THE WORLD'S GREAT SPEECHES (Fourth Enlarged Edition), Lewis Copeland, Lawrence W. Lamm, and Stephen J. McKenna. Nearly 300 speeches provide public speakers with a wealth of updated quotes and inspiration–from Pericles' funeral oration and William Jennings Bryan's "Cross of Gold Speech" to Malcolm X's powerful words on the Black Revolution and Earl of Spenser's tribute to his sister, Diana, Princess of Wales. 944pp. 5⅜ x 8⅜. 40903-1

THE BOOK OF THE SWORD, Sir Richard F. Burton. Great Victorian scholar/adventurer's eloquent, erudite history of the "queen of weapons"–from prehistory to early Roman Empire. Evolution and development of early swords, variations (sabre, broadsword, cutlass, scimitar, etc.), much more. 336pp. 6⅛ x 9¼. 25434-8

AUTOBIOGRAPHY: The Story of My Experiments with Truth, Mohandas K. Gandhi. Boyhood, legal studies, purification, the growth of the Satyagraha (nonviolent protest) movement. Critical, inspiring work of the man responsible for the freedom of India. 480pp. 5⅜ x 8½. (Available in U.S. only.) 24593-4

CELTIC MYTHS AND LEGENDS, T. W. Rolleston. Masterful retelling of Irish and Welsh stories and tales. Cuchulain, King Arthur, Deirdre, the Grail, many more. First paperback edition. 58 full-page illustrations. 512pp. 5⅜ x 8½. 26507-2

THE PRINCIPLES OF PSYCHOLOGY, William James. Famous long course complete, unabridged. Stream of thought, time perception, memory, experimental methods; great work decades ahead of its time. 94 figures. 1,391pp. 5⅜ x 8½. 2-vol. set. Vol. I: 20381-6 Vol. II: 20382-4

THE WORLD AS WILL AND REPRESENTATION, Arthur Schopenhauer. Definitive English translation of Schopenhauer's life work, correcting more than 1,000 errors, omissions in earlier translations. Translated by E. F. J. Payne. Total of 1,269pp. 5⅜ x 8½. 2-vol. set. Vol. 1: 21761-2 Vol. 2: 21762-0

MAGIC AND MYSTERY IN TIBET, Madame Alexandra David-Neel. Experiences among lamas, magicians, sages, sorcerers, Bonpa wizards. A true psychic discovery. 32 illustrations. 321pp. 5⅜ x 8½. (Available in U.S. only.) 22682-4

THE EGYPTIAN BOOK OF THE DEAD, E. A. Wallis Budge. Complete reproduction of Ani's papyrus, finest ever found. Full hieroglyphic text, interlinear transliteration, word-for-word translation, smooth translation. 533pp. 6½ x 9¼. 21866-X

MATHEMATICS FOR THE NONMATHEMATICIAN, Morris Kline. Detailed, college-level treatment of mathematics in cultural and historical context, with numerous exercises. Recommended Reading Lists. Tables. Numerous figures. 641pp. 5⅜ x 8½. 24823-2

PROBABILISTIC METHODS IN THE THEORY OF STRUCTURES, Isaac Elishakoff. Well-written introduction covers the elements of the theory of probability from two or more random variables, the reliability of such multivariable structures, the theory of random function, Monte Carlo methods of treating problems incapable of exact solution, and more. Examples. 502pp. 5⅜ x 8½. 40691-1

THE RIME OF THE ANCIENT MARINER, Gustave Doré, S. T. Coleridge. Doré's finest work; 34 plates capture moods, subtleties of poem. Flawless full-size reproductions printed on facing pages with authoritative text of poem. "Beautiful. Simply beautiful."–*Publisher's Weekly*. 77pp. 9¼ x 12. 22305-1

NORTH AMERICAN INDIAN DESIGNS FOR ARTISTS AND CRAFTSPEOPLE, Eva Wilson. Over 360 authentic copyright-free designs adapted from Navajo blankets, Hopi pottery, Sioux buffalo hides, more. Geometrics, symbolic figures, plant and animal motifs, etc. 128pp. 8⅜ x 11. (Not for sale in the United Kingdom.) 25341-4

SCULPTURE: Principles and Practice, Louis Slobodkin. Step-by-step approach to clay, plaster, metals, stone; classical and modern. 253 drawings, photos. 255pp. 8⅛ x 11. 22960-2

THE INFLUENCE OF SEA POWER UPON HISTORY, 1660–1783, A. T. Mahan. Influential classic of naval history and tactics still used as text in war colleges. First paperback edition. 4 maps. 24 battle plans. 640pp. 5⅜ x 8½. 25509-3

CATALOG OF DOVER BOOKS

THE STORY OF THE TITANIC AS TOLD BY ITS SURVIVORS, Jack Winocour (ed.). What it was really like. Panic, despair, shocking inefficiency, and a little heroism. More thrilling than any fictional account. 26 illustrations. 320pp. 5⅜ x 8½. 20610-6

FAIRY AND FOLK TALES OF THE IRISH PEASANTRY, William Butler Yeats (ed.). Treasury of 64 tales from the twilight world of Celtic myth and legend: "The Soul Cages," "The Kildare Pooka," "King O'Toole and his Goose," many more. Introduction and Notes by W. B. Yeats. 352pp. 5⅜ x 8½. 26941-8

BUDDHIST MAHAYANA TEXTS, E. B. Cowell and others (eds.). Superb, accurate translations of basic documents in Mahayana Buddhism, highly important in history of religions. The Buddha-karita of Asvaghosha, Larger Sukhavativyuha, more. 448pp. 5⅜ x 8½. 25552-2

ONE TWO THREE . . . INFINITY: Facts and Speculations of Science, George Gamow. Great physicist's fascinating, readable overview of contemporary science: number theory, relativity, fourth dimension, entropy, genes, atomic structure, much more. 128 illustrations. Index. 352pp. 5⅜ x 8½. 25664-2

EXPERIMENTATION AND MEASUREMENT, W. J. Youden. Introductory manual explains laws of measurement in simple terms and offers tips for achieving accuracy and minimizing errors. Mathematics of measurement, use of instruments, experimenting with machines. 1994 edition. Foreword. Preface. Introduction. Epilogue. Selected Readings. Glossary. Index. Tables and figures. 128pp. 5⅜ x 8½. 40451-X

DALÍ ON MODERN ART: The Cuckolds of Antiquated Modern Art, Salvador Dalí. Influential painter skewers modern art and its practitioners. Outrageous evaluations of Picasso, Cézanne, Turner, more. 15 renderings of paintings discussed. 44 calligraphic decorations by Dalí. 96pp. 5⅜ x 8½. (Available in U.S. only.) 29220-7

ANTIQUE PLAYING CARDS: A Pictorial History, Henry René D'Allemagne. Over 900 elaborate, decorative images from rare playing cards (14th–20th centuries): Bacchus, death, dancing dogs, hunting scenes, royal coats of arms, players cheating, much more. 96pp. 9¼ x 12¼. 29265-7

MAKING FURNITURE MASTERPIECES: 30 Projects with Measured Drawings, Franklin H. Gottshall. Step-by-step instructions, illustrations for constructing handsome, useful pieces, among them a Sheraton desk, Chippendale chair, Spanish desk, Queen Anne table and a William and Mary dressing mirror. 224pp. 8⅛ x 11¼. 29338-6

THE FOSSIL BOOK: A Record of Prehistoric Life, Patricia V. Rich et al. Profusely illustrated definitive guide covers everything from single-celled organisms and dinosaurs to birds and mammals and the interplay between climate and man. Over 1,500 illustrations. 760pp. 7½ x 10⅞. 29371-8

Paperbound unless otherwise indicated. Available at your book dealer, online at **www.doverpublications.com**, or by writing to Dept. GI, Dover Publications, Inc., 31 East 2nd Street, Mineola, NY 11501. For current price information or for free catalogues (please indicate field of interest), write to Dover Publications or log on to **www.doverpublications.com** and see every Dover book in print. Dover publishes more than 500 books each year on science, elementary and advanced mathematics, biology, music, art, literary history, social sciences, and other areas.

CPSIA information can be obtained
at www.ICGtesting.com
Printed in the USA
BVHW090312060822
643962BV00008B/327

9 780486 425559

VISION
IN
ELEMENTARY
MATHEMATICS

W. W. Sawyer

DOVER PUBLICATIONS, INC.
Mineola, New York

Bibliographical Note

This Dover edition, first published in 2003, is an unabridged reprint of the work originally published by Penguin Books Ltd., Hammondsworth, Middlesex, England, in 1964.

Library of Congress Cataloging-in-Publication Data

Sawyer, W. W. (Walter Warwick), 1911-
    Vision in elementary mathematics / W. W. Sawyer.
        p. cm.
    Originally published: Hammondsworth, Middlesex, England ; Baltimore, Md. : Penguin Books, 1964, in series: Introducing mathematics ; 1.
    ISBN-13: 978-0-486-42555-9
    ISBN-10: 0-486-42555-X
    1. Mathematics—Study and teaching (Elementary) I. Title.

QA135.6 .S39 2002
372.7—dc21

2002034825

www.doverpublications.com

Contents